RESEARCH EDUCATION SERVICE
M M F

M M F

EDUCATION SERVICE RESEARCH

MARSHFIELD MEDICAL RESEARCH FOUNDATION 1959 1989

MARSHFIELD MEDICAL RESEARCH FOUNDATION

MARSHFIELD CLINIC®
Research Foundation

Marshfield Clinic®
Research Foundation

TITLE

A Fifty Year History of Marshfield Clinic Research Foundation

AUTHOR

Joseph J. Mazza

Marshfield Clinic Research Foundation
www.marshfieldclinic.org/research

ISBN: 978-0-9815707-1-6

5558-000

Dedication

The book is dedicated to those individuals, who in the early years had the vision of complementing Marshfield Clinic with a center for medical research.

Joseph J. Mazza M.D.

About The Author

Dr. Joseph J. Mazza is an emeritus physician who is currently a senior research scientist at the Marshfield Clinic Research Foundation. He is a Hematologist/Oncologist who joined The Clinic in 1970 and practiced his specialty for 36 years before joining the Research Foundation. He is a Master in the American College of Physicians and served as Director of Medical Education and Assistant Director of the Internal Medicine residency program and Director of the Med-Peds residency program at Marshfield Clinic. He is on four editorial boards of medical journals, has published more than 130 scientific articles in peer review medical journals, several book chapters and is the Editor of The Manual of Clinical Hematology. His more recent publications include a History of The National Farm Medicine Center, an integral part of the Research Foundation, published in 2006.

He lives in Marshfield with his wife Ginny and they have four children.

A FIFTY YEAR HISTORY OF

Marshfield Clinic

RESEARCH FOUNDATION

By Joseph J. Mazza, M.D.

1959 — 2009

Table of Contents

Preface
6

Introduction
7

CHAPTER 1
The Early Years
8

CHAPTER 2
The Decades of the 70s and 80s:
A Time of Growth and Expansion
22

CHAPTER 3
The Era of New Technology
36

CHAPTER 4
The New Millennium
44

Conclusion
55

Preface

The idea of committing resources to research endeavors by the Marshfield Clinic did not begin in the 1950s as I intimate in my narrative, but rather in the early years of the Clinic's evolution. The fledgling clinic of six physicians, as far back as 1922, with the collaboration of the Sisters of the Sorrowful Mother, established a clinical laboratory which complemented some of the research aspirations of the group. Through the subsequent decades during which time the Clinic continued to add new physicians and worked collaboratively with the Sisters of The Sorrowful Mother at Saint Joseph's Hospital, the Clinic and hospital became a referral center for physicians throughout Northern and Central Wisconsin. Many of the patients came from the numerous dairy farms in the adjacent counties and the well established lumber industry that was supplying the pine lumber for homes as the country was moving west in the early 1900s. Trauma and sickness were common in the lumber camps and on the farms, and Marshfield became a hub for medical care. The group aspired to study and publish many of the unique maladies noted in this patient population and attract elite physicians. As a result, a medical library was established in the Clinic building which was a testament to the group's commitment to medical education. Members were asked to give lectures and conferences throughout the state and individuals took leave to travel abroad to learn new skills and treatments.

I have tried to capture the events and occurrences that contributed to the growth and success of the Marshfield Clinic Research Foundation over the past 50 years. I have been fortunate and grateful to have had the archives of both the Clinic and Foundation upon which to draw and a trove of priceless photographs extending back to the 20s and 30s when the burgeoning Clinic was beginning to coalesce, as well as firsthand recollections of the many individuals I have interviewed. The addition of the photos of the Foundation's early years to the manuscript was a wonderful way for me to complement and embellish its history. They depict the Foundation's ever changing environment and the individuals who were part of that evolution. Because of the frequent changes in the Foundation's name, it is simply referred to as the Foundation or Research Foundation to avoid confusion. There likely are people and events that will not be found in this history, and I have had to prioritize using my own judgment and my 40 years of being an integral part of the Foundation. For ease of reading, the research and education events are presented in chronologic order following a timeline of history. However, it is important to realize that the history really began with the establishment of the Marshfield Clinic in 1916.

— Joseph J. Mazza, M.D.

Introduction

Marshfield Clinic was founded by six Marshfield physicians including (from left) William Hipke, M.D.; Karl Doege, M.D.; Victor Mason, M.D.; Walter Sexton, M.D.; H.H. Milbee, M.D.; and Roy Potter, M.D. This painting, by Shirley Thompson, is displayed in the Clinic cafeteria.

Who would have thought over 95 years ago that in rural central Wisconsin a small group practice would evolve into a huge medical network of more than 800 physicians and more than 55 regional locations complete with its own world class research facility? Located between the Soo and Great Northern railroad lines, and established largely on the escalating lumber needs of the growing country, Marshfield seemed an unlikely community to attract physicians and scientists. However, the commitment and vision of a handful of physicians came to fruition in the passing decades with the acquisition of individuals of similar values and dreams.

In 1959, thanks to the persistence and vision of a German immigrant physician, Stephan Epstein, M.D., a fledgling Research Foundation was established to complement the growing Clinic of 35 physicians. Finances and facilities loomed as major obstacles, but these too, would be surmounted. Doubters became believers, and the community rallied to support the endeavor. It did not take long before recognition came to the Foundation through its published research, which led to financial support from outside institutions and agencies. Its early focus on farmers and their families grew into a center of excellence, the National Farm Medicine Center, and establishment of the Greater Marshfield Community Health Plan. Both would provide resources and health care for the many citizens of central and northern Wisconsin.

During the 1990s the Foundation became a reference center for genetics research and established a reputation comparable to prestigious centers throughout the country. Within the past decade the development of the Personalized Medicine Research Project established a repository of DNA specimens from approximately 20,000 individuals that serves as an invaluable resource for continued molecular and genetics research. In addition, the establishment of a Biomedical Informatics Research Center as a major component of the Foundation has led to a growing center for computational science that has attracted both domestic and international scientists. These two centers have set the course for growth and research for coming generations.

It is heartwarming to see that some of those individuals who have played a significant role in the success of the Foundation over the past 50 years are still with us. It is to them that we owe our gratitude and thanks, and it is they who provide the role models for us to emulate as we continue their legacy.

CHAPTER 1

The Early Years

Marshfield Clinic, established in 1916, quickly became a reputable center of excellence in the area and rapidly expanded its resources when it joined with the Sisters of the Sorrowful Mother, who had already established a hospital in Marshfield to provide care for those living in central and northern Wisconsin. A testament to this successful venture was the continued growth of the Clinic through the years of the Great Depression and the pre- and post-war decades.

ABOVE: The multiple examining rooms that line the hall in the early clinic.

OPPOSITE PAGE: The earliest days of the Marshfield Clinic on Central Ave Marshfield.

Arguably one of the most important decisions in the history of the Clinic, and indeed for the city of Marshfield, was the establishment of Marshfield Medical Research Foundation. In 1957 there were approximately 30 physicians on the staff of Marshfield Clinic, several with specialty training, who were attracted to the spectrum of medical care being rendered to patients in this largely rural community. The idea of a research facility as a component of the Clinic was first suggested and discussed at the home of Stephan Epstein, M.D., by a few physicians with remarkable instincts and vision. It was a novel, audacious idea, but the group thought that the Clinic had achieved enough of a reputation and level of sophistication attributable to its staff physicians to consider adding a formal research and education component. Practical issues were raised: how would the facility be financed; how would they acquire the researchers for such a facility; where would it be housed; and what would be the role of the Clinic physicians in supporting such an endeavor? The instigator of this concept was Dr. Epstein, a dermatologist who joined the Clinic in 1936 with an academic background emanating from his medical education in Germany. Dr. Epstein's undaunted pursuit to establish a research center and create a more academic image for the Clinic soon gained support in the group with Lee Schloesser, M.D.; Dean Emanuel, M.D.; Russell Lewis, M.D.; Ben Lawton, M.D.; and Frederick "Fritz" Wenzel all favoring the idea. After making application in early 1959, a charter was granted to Marshfield Clinic for the establishment of Marshfield

It was a novel, audacious idea.

Saint Joseph's Hospital on St. Joseph's Ave. in the early 1900s. It was at that time one of the largest hospitals in north central Wisconsin.

Clinic Center for Research and Education in September 1959 by the state of Wisconsin. At that time, just as the idea of the research Foundation was becoming a reality and a loose agenda was being crafted by the group, Dr. Epstein sent a letter to Dr. Tom Rice, a practicing obstetrician at the Clinic. He attempted to exploit the Clinic's "men's club" activities to acquire seed money for the fledgling Foundation by suggesting that 10 percent of the weekly poker pot be donated to this worthy cause!

Shortly after the charter was issued, a small brown residence on 6th Street was purchased to be the first home of the Research Foundation. This facility actually housed some animals being used in research projects, and it was Dr. Epstein's wife who was primarily involved in caring for additional animals at the St. Joseph's Hospital facility. The day after the charter was issued, the first slate of officers was elected and consisted of Dr. Epstein as president, Dr. Schloesser as vice president, Fritz Wenzel as secretary and Floyd Detert as treasurer. Wenzel had already established himself as a leader in the Clinic laboratory and Detert was at that time Clinic administrator. Dr. Schloesser was an internist and hematologist who had completed his specialty training at the University of Wisconsin, joining the Clinic as its first hematologist.

Stephan Epstein M.D., a dermatologist who joined the Clinic in 1936, became the organizing force for the research component to today's Clinic.

In October 1959 the total assets of the Foundation were reported as $460.27.

In early January 1960 a trip to Washington, D.C., was arranged by Congressman Melvin R. Laird to the National Institutes of Health (NIH) to visit various laboratories and to discuss research grants with the appropriate people at that institution. That train trip to Washington, D.C., included Drs. Epstein, Schloesser, Emanuel and Gail Williams, as well as Fritz Wenzel. At the February 1960 Foundation meeting, changes in the articles of incorporation and bylaws were submitted for approval. At the annual meeting of the Foundation in April 1960, Dr. Epstein pointed to three issues that needed to be addressed: organization, adequate facilities, and money. Noting the Foundation's legal independence, he expressed hope that the Foundation would also be financially independent within several years. The treasurer's report at that

Form 103—Sec. State.
Certificate of Incorporation

United States of America
State of Wisconsin
Department of State

M-10428

To All to Whom These Presents Shall Come:

The undersigned, as Secretary of State of the State of Wisconsin, hereby certifies that, on the 10th *day of* September *, 195 9, Articles of Incorporation were filed in my office, under the provisions of Chapter 181 of the Wisconsin Statutes, for the following corporation formed*

WITHOUT STOCK AND NOT FOR PROFIT:

Name: MARSHFIELD CLINIC FOUNDATION, FOR MEDICAL RESEARCH AND EDUCATION, INC.

Principal Office: Marshfield, Wisconsin.

Purposes for which organized: to aid, engage in, conduct and foster medical, surgical and scientific research and investigation in the broadest and most liberal manner and to apply the knowledge derived therefrom to the causes, prevention, relief and therapy of human diseases and the benefit of mankind in general; etc.

I hereby certify that a certificate has been filed in my office to the effect that a duplicate of said Articles, bearing my certificate, was recorded in the office of the Register of Deeds of Wood *County, Wisconsin, on the* 14th *day of* September *, 1959 .*

THEREFORE, *The State of Wisconsin does hereby grant unto the said corporation the powers and privileges conferred by the Wisconsin Statutes for the purposes stated and in accordance with said Articles.*

In Witness Whereof, I have hereunto set my hand and affixed my official seal, at the Capitol, in the City of Madison, this 15th *day of* September *A. D. 195* 9 .

Secretary of State

LEFT: Marshfield Research Foundation is officially incorporated September 14th, 1959.

BOTTOM LEFT: The Little Brown House, located on 6th Street in downtown Marshfield, served as the Foundation's first home.

BOTTOM RIGHT: The Dog Lab, located adjacent to St. Joseph's Hospital where Dr. Sautter performed cardiovascular surgery on dogs and horses using his newly designed heart-lung by-pass apparatus.

Dean Emanuel M.D., received the 1st federally funded grant to the Research Foundation in 1960 and became a noted international researcher for his subsequent research on farmer's lung and other environmental pulmonary diseases. His accomplishments set the tone for the Foundation for the decades to come.

Dr. Richard (Bud) Sautter, was a cardiovascular surgeon, researcher and director of the Medical Research Foundation and Medical Education.

meeting showed a balance of $1,751. Fundraising was discussed at length and a fund drive committee was appointed, but it was never the intent of the founders that the Foundation would have an eleemosynary existence. In August 1960, Dr. Emanuel, who joined the Clinic in 1958, received notification of a five-year, $41,219 grant for research on chronic pulmonary disease occurring in farmers that was coined "Farmer's Lung." It was the Foundation's first federal grant.

In September 1960, the Foundation applied for recognition by the IRS as a tax free corporation and by December that year the treasurer's report showed a balance of $6,600. Also during that year, Gail Williams, M.D., a general surgeon, established the first tumor registry in Wisconsin. In February 1961, Congressman Laird announced a $31,430 grant to Dr. Epstein from the NIH for studying drug-related delayed photo allergic sensitivity in animals. This was the second NIH grant in six months awarded to the Foundation and in March of that year, the IRS approved the Foundation's request for tax exempt status. These events were a huge boost of confidence to the young Foundation and generated more interest among Clinic staff for the entire venture.

Later that year, in June 1961, the Foundation began discussing the possibility of inviting local businessmen and community leaders to serve on its Board of Directors. Fundraising within the Clinic also began with the canvassing of Clinic physicians for contributions to support the Foundation, and was met with a favorable response.

In August 1961, Richard "Bud" Sautter, M.D., a cardiologist and thoracic surgeon from the University of Iowa, joined the Clinic and quickly became involved in research at the Foundation. He is credited with performing the first cardiac surgery at St. Joseph's Hospital in June 1962. He would later be awarded multiple large grants for his work in thoracic and cardiac surgery research. He would also become director of the Research Foundation.

During that year, five papers were published and an additional five on pulmonary emboli were accepted for publication. Fifteen research projects were completed and educational activities involving the Clinic staff, pre-med and medical students and a scientific exchange program involving other clinics in Wisconsin were established. This was a formal arrangement of an educational exchange program with other large clinics in Wisconsin, e.g., Gundersen, Monroe, Midelfort, and the Duluth and St. Louis Park clinics

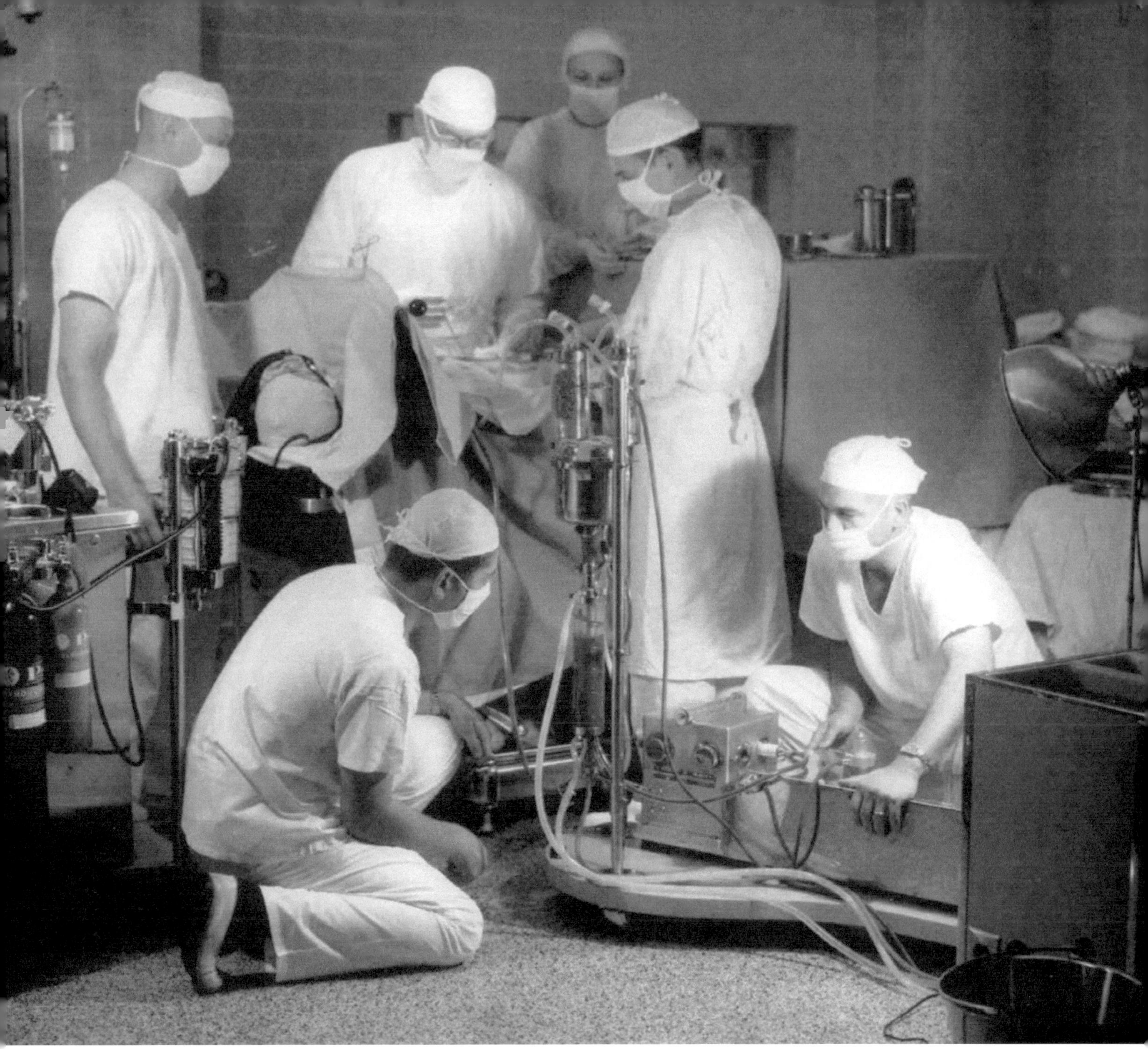

in Minnesota. This became an annual event where research projects were exchanged and openly discussed. Saturday morning conferences held at the Hospital were a weekly event with excellent staff attendance and became both a scientific and social gathering.

In February 1962, the first report of Maple Bark Disease was published by Dr. Emanuel, Dr. Lawton and Mr. Fritz Wenzel. Their research would later result in multiple grants and contracts to pursue this investigation along with other chronic respiratory diseases occurring in farmers and other rural workers. During that year, cooperation between the Foundation and St. Joseph's Hospital provided the Foundation with additional laboratory space, and the establishment of an advisory board to the Foundation was

Clockwise from top left: Dr. Warren Holtey, Dr. Richard Sautter, Bernadette Wayerski, Bob Payne, LeRoy Ferries, John Mattson. Dr. Sautter, with his Cardiac Surgery team, performing surgery using his newly designed heart-lung cardiac by-pass apparatus at St. Joseph's Hospital.

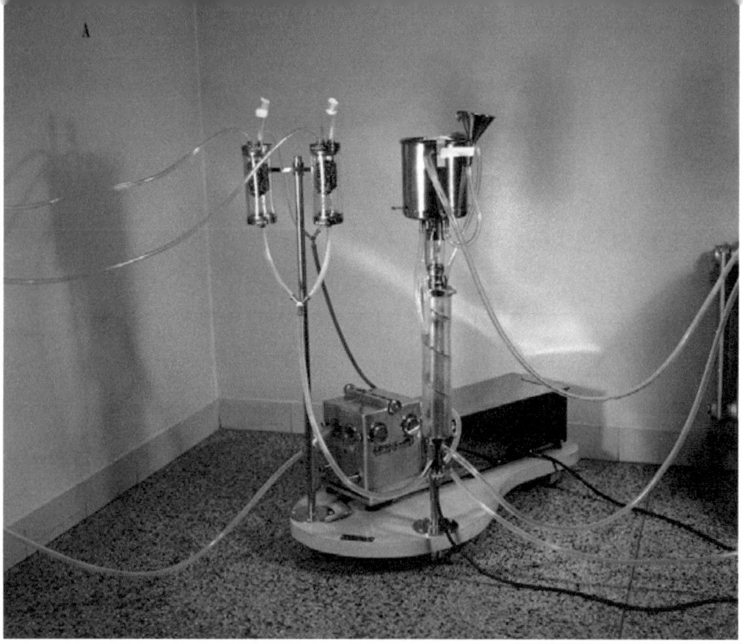

authorized. Officers elected at the annual meeting that year were: Dr. Epstein, president; Dr. Emanuel, vice president; Fritz Wenzel, secretary; and Floyd Detert, treasurer. Drs. Lewis and Schloesser were elected to the Board of Directors of the Foundation. In May of that year, there was a very successful community fund drive that involved business and community leaders and led to the establishment of the advisory board consisting of Leo Miller, Don Weber, John Figi, Lawrence Lee, Mark Hansen, Ed Zappen and Tony Knott. These were community leaders who would remain fully supportive of the Research Foundation and its future plans. During that fund drive, community members donated approximately $11,000 to the Foundation. The highlight of the year was an NIH grant for the study of Maple Bark Disease that was awarded to Dr. Emanuel as principal investigator. A second grant for Maple Bark Disease was made in December by the Owens of Illinois paper company.

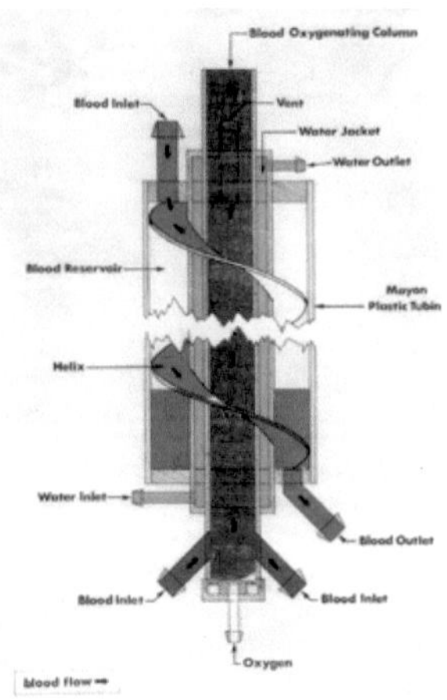

Dr. Sautter's heart-lung apparatus that enables cardiac surgery to be performed while bypassing the heart. The machine became the prototype for future designs for the same purpose.

In April of 1963 Leo Miller and Tom Felion, then administrator of St. Joseph's Hospital, were elected to the Board of Directors of the Foundation. This marked the first time that lay individuals not employed at the Clinic were elected to serve on the Board.

In March 1964, Drs. Sautter, Emanuel and Fritz Wenzel applied for and received a $10,000 grant from the Wisconsin Heart Association to study reflex circulation. Later that year the directors met and approved a contract for Duane Tewksbury, Ph.D., to join the Foundation as its first research scientist. Dr. Tewksbury had completed his studies at the University of Wisconsin and was interested in pursuing his work on hypertension at the Foundation. Dr. Tewksbury was a research biochemist and was specifically interested in the role of the renin-angiotensin system in controlling blood pressure. He would later achieve national recognition for his research on angiotensinogen and its role in hypertension.

At that time, Dr. Sautter received recognition for his pioneering work on the oxygenating apparatus that could be used in cases of heart-lung surgery, and Fritz Wenzel and Dr. Emanuel reported isolation of the first mold-like bacteria "micropolysporafaeni" as the agent causing Farmer's Lung.

In summary, 1964 was a banner year for the Foundation. In addition to all the activities and accomplishments alluded to above, there were 15 projects and six major grants received and 31 papers published or accepted for publication. Dr. Sautter's heart-lung device made Newsweek magazine, and Dr. Gail Williams reported the first five-year results of the cancer registry project that he began in 1959. Parenthetically, Dr. Sautter's new heat exchange/oxygenator/reservoir for extracorporeal circulation was presented at the American College of Surgeons meeting, as well as at other venues throughout the country including the State Medical Society meetings in Wisconsin. He worked tirelessly using laboratory animals to perfect his design of the apparatus with the able assistance of Dr. Emanuel, Bob Biechler and Fritz Wenzel. The machine allowed cardiac bypass while oxygenating the blood outside the body. It would serve as the template for the development of more modern technology so that cardiac bypass surgery would become a common, safe procedure.

Fritz Wenzel, Researcher, Director of the Medical Research Foundation and CEO of Marshfield Clinic was a major force in the growth and success of the early Foundation.

In 1964, Wenzel became the first formal director of the Foundation.

Dr. Epstein submitted a letter of resignation to the board, but was persuaded to continue as president until the next annual meeting in 1965. At the Foundation's first meeting in 1965, Dr. Epstein resigned and delivered his final remarks as president by thanking the many people who were involved in the Foundation's success and for their scientific achievements. He then turned the reins of the Foundation over to Dr. Emanuel who was elected by the board as its president. Wenzel became vice president, Frank Lohrenz, M.D., became secretary and Detert remained treasurer. In early 1965 another trip to NIH in Washington, D.C., was arranged by Congressman Laird to discuss research programs with grant administration officers in the interest of facilitating extramural grant funding to support the research projects at the Foundation.

Dr. Frank Lohrenz M.D. endocrinologist and clinical investigator received a federally funded grant to establish a Physician Assistant Program (PA) at the Clinic.

Scientific exhibits representing the Foundation, were shown at Wisconsin Medical Society meetings by a number of clinicians who had completed their research projects. Later that year, a new three-year grant of $47,000 for continued research of Farmer's Lung was announced, and Dr. Tewksbury accepted his first grant from the Kramers Urban Pharmaceutical Co. in Milwaukee. This would be one of a series of grants by this company extended to Dr. Tewksbury for his work on the isolation of plasma proteins related to hypertension.

Nineteen hundred sixty-five was another banner year for the Foundation. Multiple specialty conferences were held throughout the year, and the Clinic exchange program continued. Educational programs for high school and undergraduate students in Wisconsin were added to the Foundation's educational activities. Twenty-three projects were completed or in progress

that year, of which nine were externally funded, and 29 papers were published or accepted for publication. In December of that year, the Foundation accepted an $8,000 donation from the Steve J. Miller Foundation, the largest single, private gift to the Foundation to that date.

In 1966, Ron Roberts, Ph.D., became the second scientist to join the staff of the Research Foundation after completing his academic studies at the University of Minnesota as a protein chemist. Dr. Roberts had a deep interest and expertise in plasma proteins and later published numerous articles on plasma protein alterations in chronic diseases and blood clotting and bleeding disorders associated with protein abnormalities.

ABOVE: Duane Tewksbury, Ph.D., Fritz Wenzel and Dean Emanuel, M.D.

BELOW: John Mattson taking care of the animals in the animal quarters located in the basement of the Little Brown House, the foundation's first home.

In January 1966 the directors elected Dr. Frank Lohrenz president, Wenzel vice president, Dr. Gerald Porter secretary, and Detert remained as treasurer. In May, Thomas Nikolai, M.D., put in a successful request for a major grant to study inherited disorders of thyroid-binding globulin and Dr. Emanuel and colleagues received a seven-year, $350,000 grant from NIH to study coronary heart disease. This was called the Coronary Drug Project and its focus was on decreasing cholesterol and the incidence of coronary heart disease. The study was conducted in collaboration with multiple institutions throughout the country. Five different drugs were being tested in the study, comparing the results of each being administered over an extended period of time.

A proposal was made to the board by Dr. Russ Lewis that four members of the lay advisory board be added to the Board of Directors of the Foundation.

During that year, 30 presentations by Foundation scientists and Clinic physicians were given at state and national meetings. Drs. Roberts and Tewksbury began teaching biochemistry at St. Joseph's Hospital School of Nursing and School of Medical Technology, and Dr. Sautter was granted a patent for his cardio-pulmonary bypass apparatus. The total income of the Foundation that year, plus the balance, was $101,780.52.

In 1967, James S. Vedder, M.D., paid tribute to Dr. Epstein as the father of the Medical Research Foundation. Dr. Lohrenz was re-elected president, Dr. Emanuel became vice president, Dr. Nikolai was elected secretary and Will Gee, an executive at Weyerhaeuser Co., was named treasurer. William Schorr, M.D., a Clinic dermatologist, was elected to the executive committee.

Dr. Roberts received an $18,000 grant from the American Cancer Society to study myeloma proteins. Other grants that year included the Wisconsin Heart Association providing $10,000 for the study of pulmonary emboli and the renin-angiotensinogen system in hypertension. Another grant of $79,000 from the Wisconsin Regional Medical Program was provided for the study of pulmonary embolism with Drs. Sautter, Emanuel and Fritz Wenzel as co-principal investigators. Their work involved the administration of urokinase for the treatment of pulmonary emboli, and this research would be presented at the American College of Cardiologists meeting in Washington, D.C., that year.

"In December 1965 the Foundation accepted an $8,000 donation, the largest single, private gift to that date."

In 1967, there were 24 projects in progress and 13 grants awarded to the Foundation. Grant income that year was $107,223 and the total balance after expenditures was $138,216.

In 1968, Dr. Emanuel was elected president, Donald Fullerton, M.D., became vice president, Dr. Nikolai remained secretary and Will Gee continued as treasurer. Norbert Koopmans, Ph.D., then dean of the UW Marshfield campus, was elected to the executive committee. In August, Dr. Fullerton, a clinic psychiatrist, and Gary Gustafson, a social worker, proposed that the Foundation develop and underwrite a half-way house that would help patients make the transition from hospital to community. Discussions were held on leasing the old Doege Clinic building at 510 North St. Joseph's Avenue for the Foundation and by November 1968, the executive committee approved leasing the building and discussions on the half-way house continued. In December, Dr. Tewksbury accepted a second grant from Kramers Urban Pharmaceutical Company to study plasma proteins affecting cardiovascular function. The Saturday

Ron Roberts, Ph.D., Duane Tewksbury, Ph.D. and Fritz Wenzel.

In 1967, the Foundation moved to the old Doege Clinic across from St. Joseph's School of Nursing on St. Joseph's Ave.

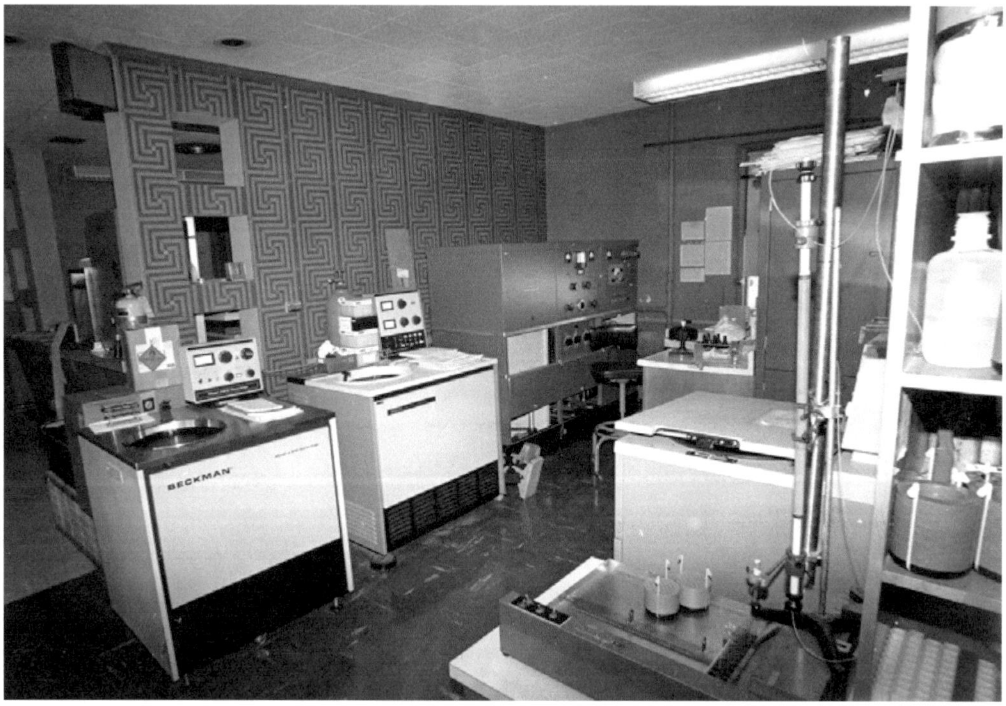

The crowded space in the early Foundation became apparent with the acquisition of the necessary equipment.

morning conferences continued weekly, and the Clinic exchange program with clinics throughout Wisconsin and Minnesota continued. Forty-three papers were published or accepted for publication that year, and Dr. Ron Roberts and Jim Travis, Ph.D., from the University of Georgia, received national recognition for their isolation of human trypsin.

On October 7, Dr. Sautter performed the first cardiac bypass procedure at St. Joseph's Hospital -- on a patient from New Lisbon, Wisconsin -- using his patented bypass apparatus that he had perfected at the Foundation. It was a landmark event that had a major impact on the Clinic and Hospital, laying the groundwork for future surgical research at the medical complex.

In 1968, Dr. Sautter performed the first cardiac bypass procedure at St. Joseph's Hospital using his patented bypass apparatus that he had perfected at the Foundation.

That year the Foundation moved to its new location in the old Doege Clinic building on North St. Joseph's Avenue that would provide the much needed additional space for the Foundation's laboratories and research aspirations, as well as administrative personnel. Major renovation and expansion to the building would soon follow.

In 1969, Dr. Dean Emanuel and Fritz Wenzel continued in their roles of president and executive director, respectively. Dr. Don Fullerton was elected vice president with Dr. Tom Nikolai secretary and John Figi treasurer. Drs. Lewis, Nikolai and Schorr were added to the Board of Directors along with Earl Nelson and James Jay, two local businessmen.

In April 1969, the executive committee approved the purchase of a house on South Maple Avenue to serve as a half-way house sponsored by the Foundation and the city council approved the purchase and rezoned the area in June. After an extensive renovation of the Maple Manor Halfway House, it received its first residents on September 1, 1969. The Foundation viewed this initiative as being responsive to its mission, which included community service.

Maple Manor was the Dept. of Psychiatry's outpatient facility for individuals requiring minimal care. It became a second home for a number of Clinic patients and functioned under the aegis of the Foundation.

In August 1969, Ed Plotka, Ph.D., accepted a contract offer from the Foundation as an endocrine physiologist, having completed his post-graduate work at Purdue University in Indiana. Dr. Plotka had a deep interest and expertise in reproductive physiology and would work closely with the members of the department of Obstetrics and Gynecology on multiple projects concerning fertility and infertility. In October of that year, Al Zimmermann

RIGHT: Dean Emanuel by the new Marshfield Medical Foundation sign.

BELOW: The Medical Library in the Clinic building which was located on Central Avenue, across from Washington Square.

OPPOSITE PAGE: Teaching and learning at the Foundation continued throughout the years. Pictured are Don Zais and Don Stoiber.

was hired jointly by the Clinic and Foundation as a medical editor to help with scientific writing and publications, and would later become manager of library services in the medical complex. Educational programs at the Clinic continued to expand and a University of Wisconsin teaching service was established with the Clinic and St. Joseph's Hospital under the direction of George Magnin, M.D. This would provide education programs for medical students from the University of Wisconsin School of Medicine, as well as an intern and internal medicine resident who would be involved in inpatient training at St. Joseph's Hospital. This would be called the "University Teaching Service." The program was largely due to the existing affiliation with the University of Wisconsin Medical School. The Clinic was already involved in taking third- and fourth-year medical students, along with the preceptorship program that involved a number of Clinic physicians who were mentoring students at various levels of their education. In addition, Rheumatology, Ophthalmology, Pediatrics and General Surgery residents from the University of Wisconsin also rotated through the Clinic's specialty departments. The Clinic and Foundation later developed an international program with medical schools in Scotland and Germany and began taking their students for two-month rotations in the medical complex. This was a very popular program for those students and continued for years after.

In 1969, 36 projects were in progress and 26 grants were awarded or applied for. The grants and contracts amounted to approximately $255,600 and year-end assets were just under $221,000. The education programs expanded and the Clinic's exchange program continued. Administratively, the Foundation submitted the first budget in its 11 year history to the executive committee and the total liabilities and assets at the end of 1969 were $220,995.24.

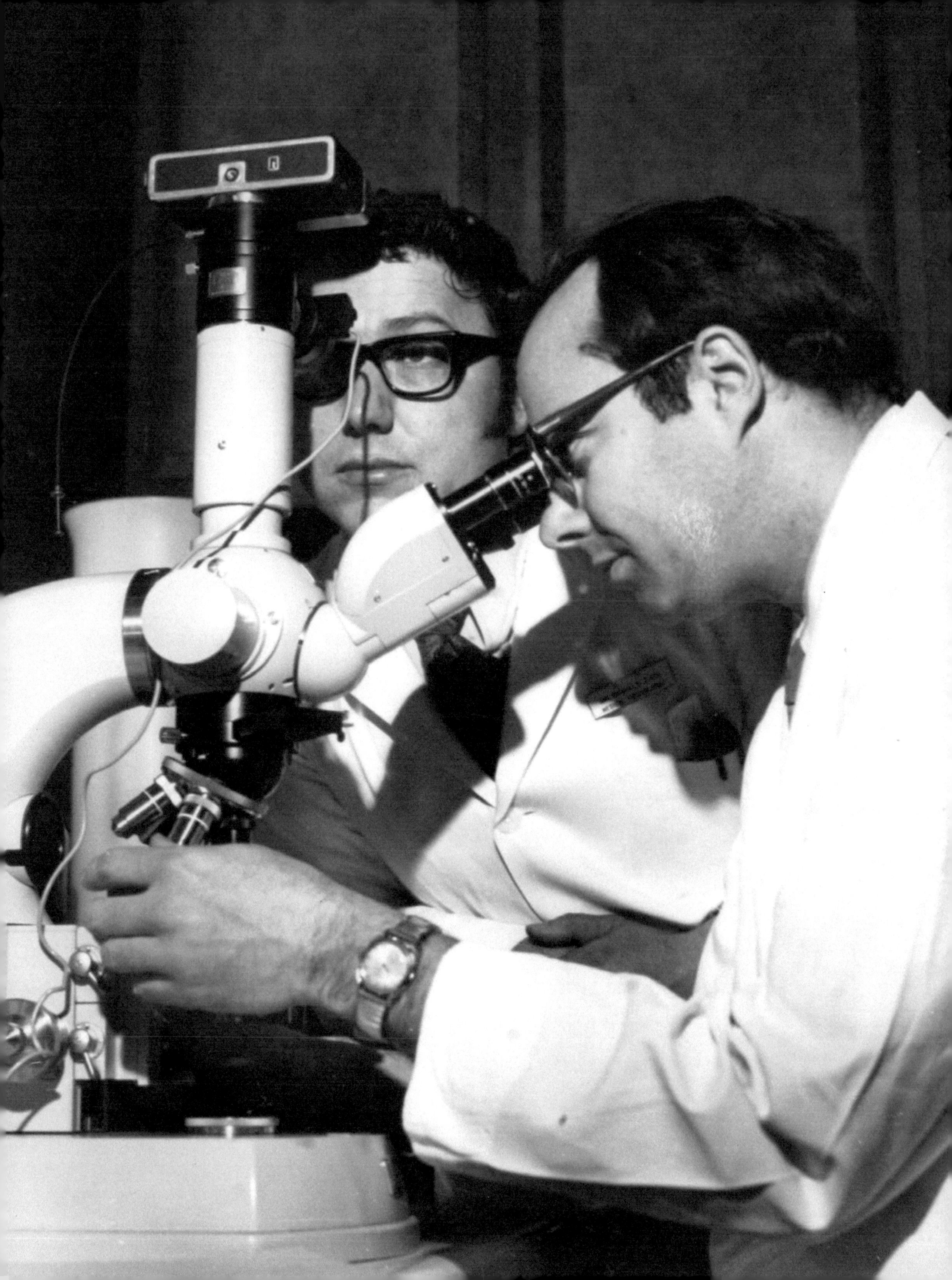

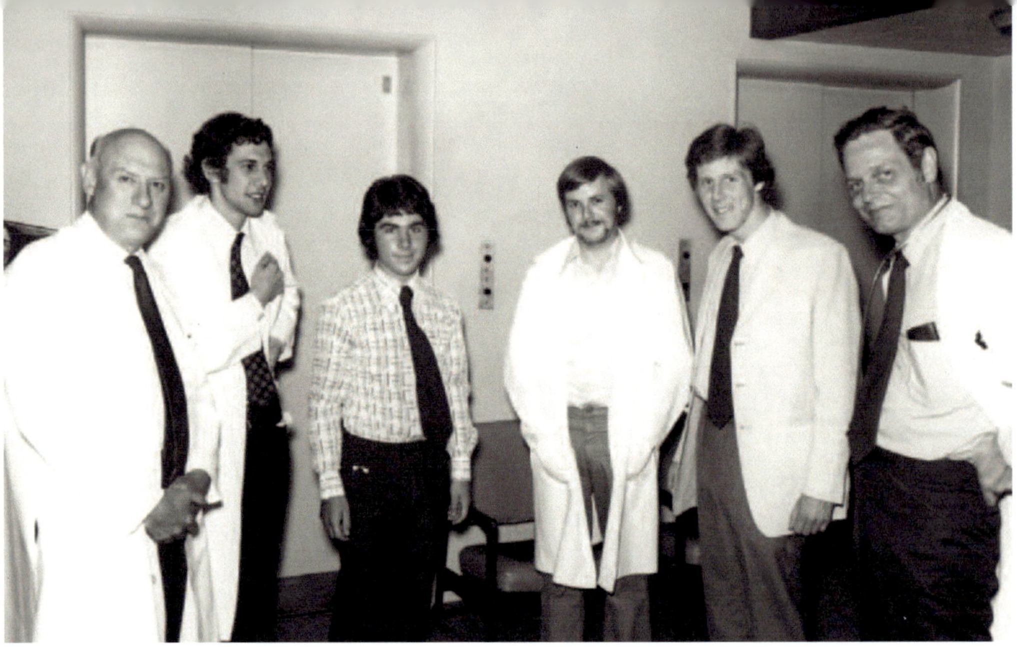

CHAPTER 2

The Decades of the 70s and 80s:

A Time of Growth and Expansion

The decade of the 70s could best be described as a period of growth and expansion. In 1973 the Foundation added a number of mobile home units to the old Doege Clinic building on North St. Joseph's Avenue to provide space for its expanding administrative staff and would later (1978) add an 8,000 square foot addition to the back of the building to accommodate the desperate need for additional laboratory space. Education programs also expanded and the University of Wisconsin Teaching Service was in place at St. Joseph's Hospital under the direction of Dr. George Magnin. There was also an increasing number of students involved in education programs throughout the medical complex.

ABOVE: Dr. George Magnin with Dr. Susan Turney, an internal medicine resident, making chart rounds at St. Joseph's Hospital.

OPPOSITE PAGE: Drs. George Magnin and Jim Manier with a group of residents.

In 1970, the officers of the Foundation were all re-elected for another term. There were 53 manuscripts published or accepted for publication and 38 different research projects in progress.

In 1971 Dr. Emanuel remained as president, Dr. Nikolai as vice president, Paul Umhoefer became secretary and Dr. Norbert Koopman was elected treasurer. That year the Clinic played host to 17 invited guest speakers, and the Clinic's exchange education program with other clinics throughout Wisconsin and Minnesota continued.

On July 1, the Maple Manor Halfway House was licensed by the State Department of Health and Social Services, and during that year the Manor served more than 30 residents. Roger Lesneski and Jerry Carson, along with social workers at the Clinic, were responsible for the success of this program, as well as Howard and Temple Beers who served as "house parents" at the Manor.

The Farmer's Lung Project continued under Drs. Emanuel, Roberts and Fritz Wenzel's direction, and a large grant was submitted to NIH for the study of chronic respiratory diseases, which would be approved and awarded the following year. This became known as the SCOR grant, identifying the Clinic as a "Specialized Center of Research" in the United States for the study of hypersensitivity lung disease. It was a five-year NIH grant that would be renewed in 1976 and again in 1981. At that time, there was virtually no data on the incidence and prevalence of Farmer's Lung among the farmers

of the Midwest, and research continued on other chronic respiratory diseases in that population. In September of that year, the north central Wisconsin Outreach Project was initiated by the Foundation through a grant from the Wisconsin Regional Medical Project. This was an important project aimed at improving the efficiency of transmitting medical information and increasing the speed of interpretation of laboratory and procedural results to patients in north central Wisconsin.

A landmark event that occurred in 1971 was the establishment of the Greater Marshfield Community Health Plan (GMCHP). Partners in the Plan were St. Joseph's Hospital and Wisconsin Blue Cross. This program provided community-wide open enrollment and community rating for all participants. The plan gained national recognition as one of the earliest health maintenance organizations in the Midwest. It was unique in that it was developed in a rural rather than urban community of Wisconsin, and grew into one of the largest prepaid health plans in the state. Dr. Russ Lewis was the director of the plan, which in 1987 became Security Health Plan (SHP).

In September of 1972 Mr. Greg Nycz joined the Research Foundation and throughout the 1970s and 80s he worked with both Clinic and Foundation leadership to assist in their vision to provide access to the innovative GMCHP for everyone in the community. He worked closely with Fritz Wenzel and Dr. Lewis in a program that integrated low-income residents into the GMCHP. Begun in March of 1974 and shepherded by Nycz, the program continues to this day. He also helped establish the first prepaid Medicaid contract in the state and later achieved the second prepaid Medicare contract that provided entrance into the GMCHP for all community residents.

That year the Physician Assistant Training program was initiated with a five-year NIH grant from the Bureau of Health Manpower. The grant was for $578,000 and would be a cooperative effort with UW Marshfield/Wood County Campus, St. Joseph's Hospital and Marshfield Clinic participating in the training program. The principals involved in this grant were Dr. Lohrenz and Robert Payne, who was a registered nurse and physician assistant working at that time at the Research Foundation. The program consisted of a robust didactic curriculum, as well as clinical experience at St. Joseph's Hospital and Marshfield Clinic. It would be the first Physician Assistant Training Program established in Wisconsin. Ten students were accepted to each class and the program operated continuously until 1980. Many members of the Clinic's staff and Foundation scientists participated in the program and served as faculty.

The scientific exchange program with the Wisconsin and Minnesota Clinics continued and the number of scientific visiting professors to the Clinic and Foundation expanded. The Foundation added James Marx, Ph.D., to its scientific staff in August 1972 after he completed his post-graduate work at the University of West Virginia, where he earned a Ph.D. in molecular biology and immunology. His graduate education and research fit in nicely with the Foundation's projects and accomplishments on chronic respiratory diseases and he quickly joined the team of scientists working on Farmer's Lung. He later became immersed in other chronic respiratory diseases and received other external grants and contracts to pursue his research.

Dr. Bill Schorr was president of the Foundation at that time, Paul Umhoefer, a local business man and ardent supporter of the Foundation, was vice president, with Dr. Duane Tewksbury secretary and Dr. Tom Nikolai as treasurer.

The increasing number of papers published and presentations listed in the education section of that year's annual report are a testament to the continuing success of the Foundation. Additionally, the records of the Foundation show that 43 research projects were in progress, 31 of which were supported by external grants or contracts.

CLINIC EMPLOYEES ENROLL IN "GMCHP"

by DONALD A. NYSTROM

On March 1st, Marshfield Clinic employees and their dependents became the first participants of the first comprehensive prepaid health care program ever organized in Wisconsin. The program is known as the "Greater Marshfield Community Health Plan."

The Clinic, St. Joseph's Hospital and Associated Hospital Service (Blue Cross) united their abilities and resources to create the unique and innovative health care plan. In prepayment plans, physicians and medical facilities combine to provide comprehensive care to a specified number of people in return for a predetermined monthly fee. Several existing prepayment plans operating in other parts of the country were studied, and the best aspects of each were carefully merged into the Marshfield Plan.

The Marshfield Plan promises to provide as broad a range of medical service and hospital service as possible with emphasis placed on the prevention of illness, early disease detection, health maintenance and alternatives to inpatient hospital services.

It is hoped that rising health care costs can be stabilized through the prepayment mechanism. A fundamental goal is to maximize the effectiveness of physicians and other health professionals by using paramedical and other personnel to the maximum of their skills.

An office for handling community enrollment in the "Greater Marshfield Community Health Plan" has been established in St. Joseph's Hospital. Representatives are on hand to answer questions and provide information to the public. Over 35,000 Marshfield area residents are eligible for membership in the plan.

Marshfield is rapidly moving ahead with the progressive new program for better health care . . . "The Greater Marshfield Community Health Plan."

We care!

The Greater Marshfield Community Health Plan provided community-wide open enrollment and community rating for all participants, which in 1987 became Security Health Plan (SHP).

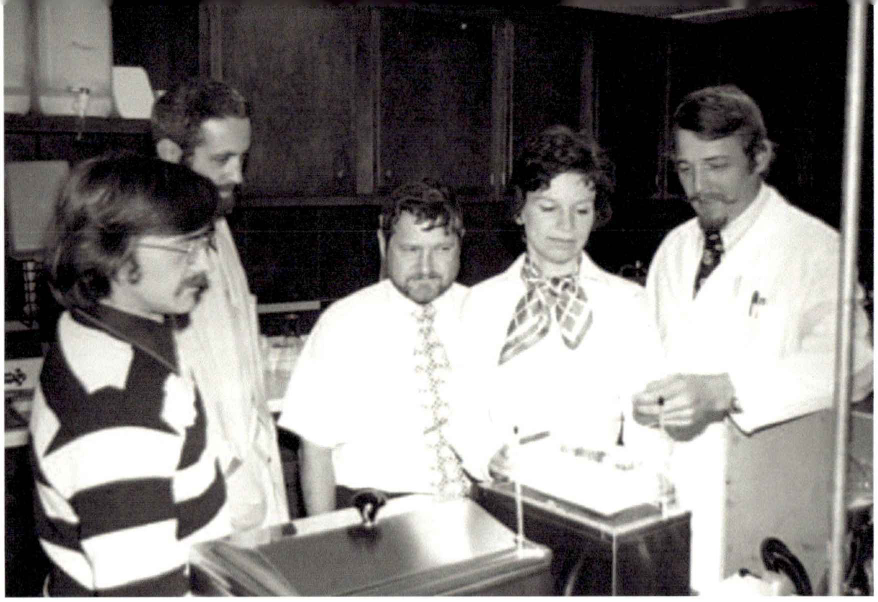

The core research staff of the Research Foundation in 1975 consisted of Drs. Marx, Plotka, Roberts, Treuhaft, and Tewksbury.

In 1973, the Medical Research Foundation was selected as one of 12 institutions to participate in a large NIH study of coronary artery disease. This was a seven-year grant which was called Coronary Artery Surgery Study (CASS), and its principal investigators were William Myers, M.D., and Dr. Sautter. The purpose of the study was to evaluate coronary artery surgery and compare it to medical treatment in prolonging life. This study would be conducted in collaboration with 11 other research institutions throughout the country. In 1973 cardiac bypass procedures were performed frequently at St. Joseph's Hospital, and it soon became apparent that Marshfield Clinic/St. Joseph was one of the leading centers in enrolling patients to the CASS study.

In June 1973, Dr. Epstein died and a tribute to his special place as the founder of Marshfield Medical Research Foundation was delivered by Drs. Schorr and Emanuel. It was Dr. Emanuel who said, "It is significant that at the time of his (Dr. Epstein's) death, the Foundation was engaged in preparing for its first major expansion. To this man, who had such wisdom, skill, insight and foresight, we shall be forever grateful."

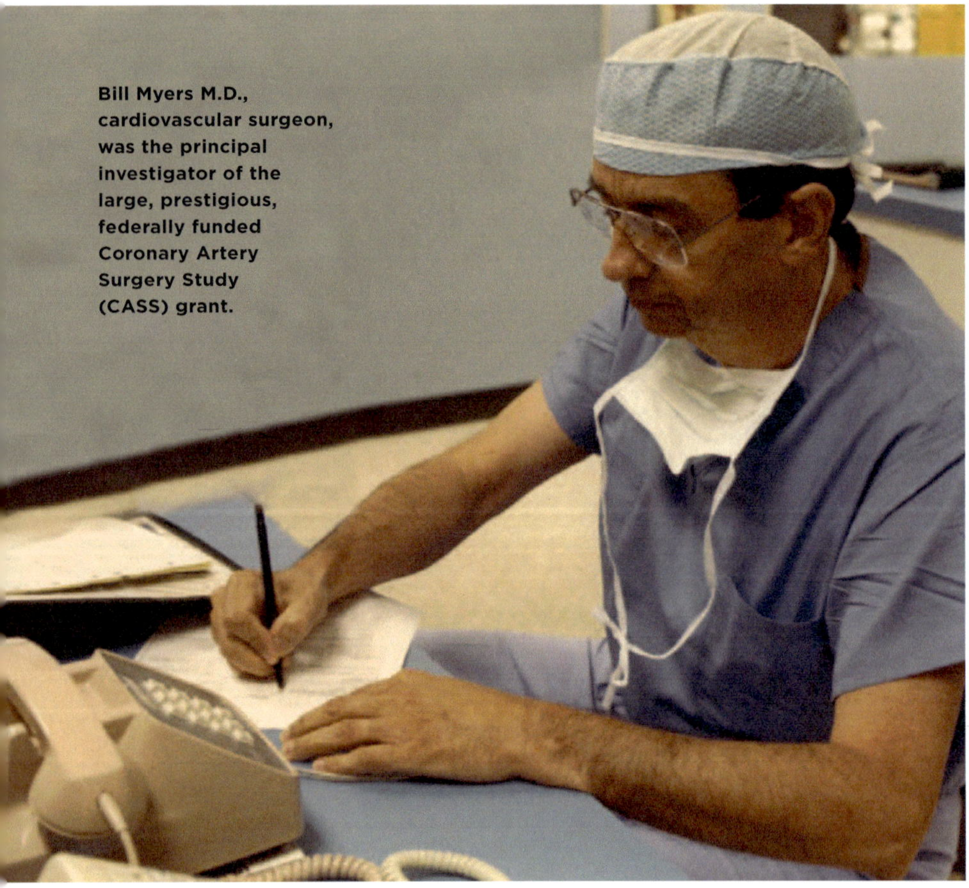

Bill Myers M.D., cardiovascular surgeon, was the principal investigator of the large, prestigious, federally funded Coronary Artery Surgery Study (CASS) grant.

By that time the Foundation had outgrown its space at the Doege building and several mobile home units had been added to provide for the expanding administrative staff. Also during that year, Dr. Tewksbury received several grants for his work on the isolation of tryptic peptides from casein and a grant from the NIH for his research on the isolation of human plasma protein fractions and their relationship to hypertension. His laboratory was very effective in isolating the array of proteins in human plasma, and his work was of great interest not only to the pharmaceutical industry, but also to NIH, and brought national recognition to the Foundation.

The Physician Assistant Training Program was off and running in 1973. Robert Intress, also a P.A. on staff, became heavily involved with the program and would later become its assistant director.

Bob Intress, PhD, was a cardiovascular physician assistant, member of staff faculty and the Assistant Director of Research and Medical Education under Dr. Sautter.

There were 45 research projects underway in 1973, 29 of which were fully funded by outside agencies. Education programs continued to expand and there were 12 internal medicine residents, seven interns and 27 medical students from the University of Wisconsin School of Medicine who received education and training at the medical complex (i.e., Marshfield Clinic and St. Joseph's Hospital) that year.

The Clinic appointed Dr. Don Fullerton as its first director of medical education in 1973 and during that year Drs. Magnin, Fullerton and Joseph Mazza, M.D., a Clinic hematologist, investigated the possibility of establishing residency training programs at Marshfield. Application was made to the American College for Graduate Medical Education (ACGME) for the establishment of a post-graduate, internal medicine residency

Mobile home structures served as temporary quarters for administrative staff while the Foundation was expanding its facility.

Physician Assistant Graduation night. Mr. David Larson receiving congratulations from Dr. Bob Intress.

training program that would be sponsored jointly by both St. Joseph's Hospital and Marshfield Clinic. The application was approved later that year and the Clinic and Hospital would select its first group of internal medicine residents who would begin their training on July 1, 1975. The program was supported by an affiliation with the University of Wisconsin School of Medicine. The accreditation of the internal medicine residency program was followed shortly thereafter by accredited residencies in general surgery and pediatrics.

Drs. George Magnin and Tom Nikolai with the first group of residents enrolled in the Internal Medicine Residency program. Drs. Romis, Fiechtner, Brosseau and Burton in 1975.

The Foundation's fund balance and assets at the end of 1973 were $258,206.

In 1974 the Foundation changed its name and became Marshfield Medical Foundation.

Don Stoiber getting familiar with the newly acquired electron microscope at the Foundation.

Dr. Tewksbury had isolated human angiotensinogen and his work was published in 1974. His purification of angiotensinogen was published two years later in 1976 and this work was supported by a grant of $72,400 from the American Heart Association. Dr. Tewksbury would receive additional grants and contracts to continue his research and would later be recognized for outstanding research by the Wisconsin Heart Association, an affiliate of the American Heart Association.

One of the highlights of 1974 was the establishment of the Family Health Center program which was developed by Dr. Russ Lewis, Fritz Wenzel and Chuck Gendron, to provide health care to low income families. In addition, the Comprehensive Child Care Center was developed to focus on children and their health care problems as separate entities. The activities of this center would serve to ensure proper interface between the interdisciplinary providers of childcare, and eliminate duplication of efforts in care and responsibility. The center would also serve as an informational resource, as well as provide essential services and become a model for other rural communities. That year the Foundation added its first electron microscope to its laboratory, and the first class of the Physician Assistant Program graduated.

Greg Nycz worked with Fritz Wenzel and Dr. Russ Lewis to launch the Family Health Center in 1974. From 1980 to 1990 he served as its Project Director; then became Executive Director of the Family Health Center Corporation in 1990 when it separated from the research division.

In 1975, Dr. Nikolai became president of the Foundation and would remain president for the next two years. The second half of the 1970s was highlighted by major expansions of both the Clinic and Research Foundation. In May 1975 the Clinic moved to its new home in the northwest part of Marshfield, adjacent to St. Joseph's Hospital. This was a newly constructed building (211,111 square feet) large enough to accommodate 150 physicians and provide ample space for the rapidly growing staff of the Clinic. Additionally, its proximity to St. Joseph's Hospital improved the efficiency of rendering care and expediting critical interventions at the hospital by Clinic physicians when necessary. The building was constructed by Marshall Erdman and Associates from Madison, and they would later construct several research buildings in the adjacent area of the expanding campus.

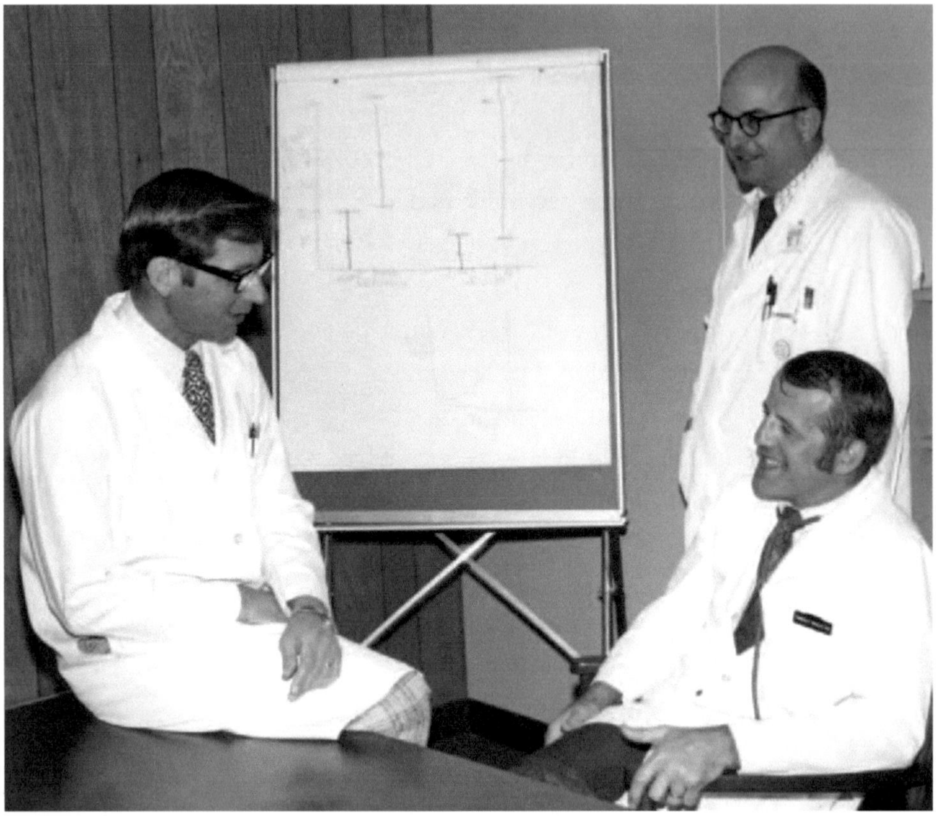

LEFT: The animal quarters during the early years of the Foundation.

TOP RIGHT: Drs. Bill Schorr, Bob Greenlaw and Tom Nikolai reviewing the drawings of the proposed new Clinic building.

RIGHT: Newly constructed Marshfield Clinic building in 1975 on the northwest side of Marshfield in close proximity to St. Joseph's Hospital.

Jim Marx Ph.D., and Ron Roberts Ph.D. collaborating on a research project at the Foundation.

In 1975, Kenneth E. Merwin was appointed as the first director of Development, a new office at the Foundation, to develop programs to generate external funding from grants and contracts and other sources to support the research projects of the Foundation. This would later lead to a larger, more formally structured "Development Department."

The Foundation continued to grow and acquire additional grants and contracts. These activities were focused on its mission to provide research, education and community service. Both clinical and basic research accomplishments were noted by the numerous publications resulting from these projects, providing the Foundation with a national reputation and opened the door for other opportunities and expansion. In 1975, Mary Treuhaft, Ph.D. joined the Foundation after completing her post-graduate work at the University of Chicago where she earned her doctorate in microbiology. She quickly became involved in identifying the microorganisms involved in chronic pulmonary diseases and hypersensitivity pneumonitis that were common in the farmers of Wisconsin. At that time the number of people employed at the Research Foundation had risen to more than 50. These included newly acquired Ph.D.s, technicians and administrative personnel.

The Farmers' Lung and related pulmonary disease studies, breast cancer research, coronary artery surgery study, angiotensinogen and its relationship to blood pressure, and other sophisticated research projects continued unabated through the decade. An additional program announced in 1976 was the planning grant from the National Cancer Institute (NCI) for a major cancer center that would enable the Foundation to bring together existing resources for detecting and managing cancer. The Foundation

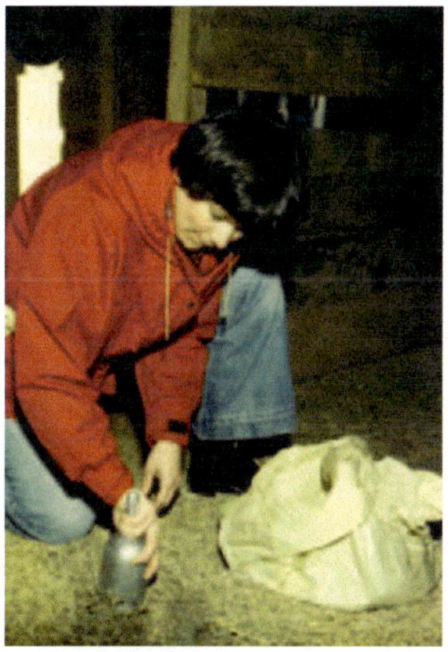

Dr. Mary Treuhaft gathering material from a local farmer's barn to study the cause of farmer's lung disease.

had been working with the Wisconsin Clinical Cancer Center in Madison and this association linked the Foundation and Clinic to the Eastern Cooperative Oncology Group (ECOG). This connection provided the oncologists at Marshfield Clinic with multiple treatment protocols and opportunities to be involved in cancer research on a national level.

In 1975, Dr. Emanuel identified the new disease, pulmonary mycotoxicosis, differentiating it from Farmers' Lung. A year later, Wenzel became executive director of the Clinic and Dr. Sautter became director of the Research Foundation the following year. Dr. Lohrenz, the director of the Physician Assistant Training Program, died in 1976. Dr. Nikolai remained president of the Research Foundation and the Foundation was named the Firm of the Year by the Marshfield Area Chamber of Commerce and Industry. In 1978, Dr. Schorr became president of the Foundation and would continue in that position for the next three years. During that year, because of the need for additional laboratory and administrative space, a large addition was added to the Foundation building, the old Doege Clinic, that provided 8,000 additional square feet and served as the Foundation's home until 1990.

National **Farm**Medicine Center®

Throughout the decade of the 80s, the Foundation continued to expand and acquire additional grants and contracts to support its many research endeavors. The SCOR grant was renewed in 1981 and additional rural health projects were initiated. A feasibility study was conducted to assess the value of developing a specific program at the Foundation that would focus on the health of the farm population in central and northern Wisconsin. This was initially referred to simply as the "Farm Center." The following year (1982) the National Farm Medicine Center became a reality and a major component of the Research Foundation. It was envisioned that this center would serve not only the medical and scientific community affiliated with the farming community, but would also reach out directly to the farming population.

There was an emphasis on cancer research during the early 1980s at the Foundation. Dr. Ed Plotka, a senior scientist at the Research Foundation, and James Hoehn, M.D., a surgical oncologist with a specific interest and expertise in breast cancer at Marshfield Clinic, worked on a project aimed at performing and improving hormonal receptor assays in patients with breast cancer. With the introduction of the hormone receptor assay they improved the ability to select patients who might respond to hormonal manipulation therapy.

In 1981, the Foundation became an affiliate member institution of the National Surgical Adjuvant Bowel and Breast Project (NSABP) which became a major component of the

Foundation's cancer research efforts in the early 80s. Dr. Hoehn was the principal investigator and major contributor of patients to this endeavor and continued in that role until his retirement in 2008.

Dr. Jim Hoehn M.D. general surgeon and principal investigator of the Clinic's participation in the National Breast Cancer study sponsored by the National Surgical Adjuvant Bowel and Breast Project (NSABP).

In 1983, the Foundation was one of 59 medical centers in the U.S. to be chosen to participate in a nationwide Community Clinical Oncology Program (CCOP) designated to provide the latest in cancer treatment to more people in their local communities. The Foundation was the only institution in Wisconsin to be awarded funds from the National Cancer Institute. The grant was for $248,166 over three years and has been renewed every 3-5 years. The most recent renewal was in 2008 for $4.6 million over five years. The principal investigators of the grant were Tarit K. Banerjee, M.D., a Marshfield Clinic oncologist; Homer Russ, M.D., a radiation oncologist; and Dr. Hoehn. Dr. Banerjee remained the grant's principal investigator until his retirement in 2008.

In addition, Dr. Marx, a senior scientist at the Research Foundation along with Dr. Banerjee, developed a clonogenic assay which was an in-vitro means of determining whether a patient's tumor was responsive or resistant to specific chemotherapeutic drugs. The assay had the potential of giving the oncologist comparable treatment advice regarding a tumor as other physicians have when treating an infectious disease with antibiotics. Dr. Marx estimated that the clonogenic assay would be, at best, only 90 percent effective in predicting resistance to a drug and about 70 percent effective in predicting sensitivity to a chemotherapeutic agent.

That same year, bone marrow transplant capabilities were established at Marshfield Clinic/St. Joseph's Hospital under the direction of William Friedenberg, M.D., hematologist; Daniel Rushing, M.D., oncologist; and Dr. Joe Mazza, hematologist. This required critical supportive assistance from the Research Foundation's immunology laboratory. Drs. Friedenberg and Mazza had spent their sabbatical time learning the technology of the procedure, Dr. Friedenberg at the Fred Hutchinson Center in Seattle and Dr. Mazza at the University of Minnesota. Both allogeneic and autologous transplants were introduced to the medical complex, and these procedures were done with the full cooperation of St. Joseph's Hospital. Patient care required a comprehensive support staff of specialists available to provide their expertise and would expand treatment options for various types of cancer that have shown favorable responses to this type of intervention.

The Radiation Oncology group continued its work as part of the Radiation Oncology Therapy Group (RTOG), another nationally recognized cooperative group involved in

acquiring research data on patients being treated with radiation therapy for their cancers. Dr. Homer Russ , who chaired the program at the Foundation, recognized the importance of being involved in such a prestigious group as well as the advantages provided for patients entering into the radiation therapy treatment protocols.

In 1983, the Heartwatch program was initiated to enhance scientific knowledge of the long-term impact of health and healthy lifestyles. The program was initiated under the direction of George Griese, M.D., pediatric cardiologist, and was conceived in response to the growing recognition that it is necessary to promote prevention and wellness programs at an early age. The program targeted cardiovascular disease, since it is the leading cause of death in the United States and Wisconsin and commonly thought of as an adult health problem. However, there is no longer any question that the roots are found in early life and the condition results from interaction of the individual's lifestyle and familial or genetic inheritance. The objectives of the program were: 1) to identify children at risk for cardiovascular disease; 2) to identify high-risk families; 3) to determine the prevalence of cardiovascular disease factors in our area; and 4) to identify interventions which could prove beneficial.

ABOVE: Dr. George Griese, pediatric cardiologist and principal investigator of the Children's Heart Watch Study.

BELOW: The kids at Our Lady of Peace Elementary School who participated in Dr. Griese's "Heart Watch Program."

With the approval and support of the Marshfield School Board, the School Committee of the parochial school system, and the very enthusiastic efforts within the individual schools, 3,720 students from grades K-12 were screened utilizing family history, nutritional habits, height, weight, blood pressure, cardio-respiratory fitness tests, and complete blood lipid determinations. All of the students and their families who took part in the program received letters detailing the analysis of their individual risk factors.

Within the Marshfield community, this Heartwatch study became a highly visible example of the Foundation's research endeavors. It is interesting to note that 25 years after Heartwatch's initiation, Laura Coleman, Ph.D., a scientist at the Research Foundation and nutritionist, was studying that same cohort of children, who are now adults, and looking again at risk factors and health care issues that have ensued over that period of time.

In 1984, there were 90 publications (or accepted for publication) submitted to the medical literature and over 150 grant and contract award projects in progress. The total liability and fund balance that year was $3,046,788. In November, the Foundation hired Ralph Mueller as the new director of Development which was a key investment in their quest to increase the amount of philanthropic contributions for the anticipated expansion of the growing Foundation.

The old Doege building was home of the Research Foundation from 1967 to 1990.

In 1985, the Foundation's name changed, this time to Marshfield Medical Research Foundation.

The latter half of the decade of the 80s was highlighted by the acquisition of new personnel and other components to the Foundation to provide additional resources to undertake projects in other disciplines of biomedical research.

Within the Marshfield community, this Heartwatch study became a highly visible example of the Foundation's research endeavors.

CHAPTER 3

The Era of New Technology

During the 1980s, biomedical technology was changing rapidly and molecular biology and genetics were coming on strong. It was Dr. Sautter's decision, with full support of the Foundation's executive committee, to acquire the resources to prepare for the future. In 1986 James Weber, with a Ph.D. in molecular genetics, joined the Research Foundation. Dr. Weber had been at the Walter Reed Institute of Research in Washington, D.C., when he was recruited by the Foundation board to head up a human genetics center at the Research Foundation.

ABOVE: The Lawton Center became a resource for Research and Education in 1990.

OPPOSITE PAGE: Dr. Jim Weber in the Genetics Research Lab going over data with Zhen Yuan Wang.

In 1987 the prestigious Gwen D. Sebold Award was established by D. David Sebold in memory of his sister, a former Marshfield Clinic employee. The purpose of the award was to recognize an outstanding individual in research and/or education within the medical complex. The award is given annually and the recipient is nominated by his/her peers.

Also during that year Peter Layde M.D., M.P.H., who came from the Centers for Disease Control and Prevention (CDC), joined the Clinic and Foundation as its first epidemiologist and established what is now called the Marshfield Epidemiologic Study Area (MESA). This is a defined geographic region that has been tracked since 1991. It consists of 24 different ZIP codes in northern and central Wisconsin with 85,000 individuals who choose to receive their health care in the Marshfield Clinic system and have a unique Clinic medical record. MESA serves as an invaluable resource to facilitate population-based research at the Foundation and it has served as a database for numerous publications and continues as a resource in the epidemiology center. The key features of MESA are its large population coverage and capture rate of events as each individual member is tracked over time. The records of these individuals registered to MESA are linked to a sophisticated electronic medical record and extensive data warehouse archives. The database provides a natural population laboratory for conducting clinical epidemiology

MESA serves as an invaluable resource to facilitate population-based research at the Foundation.

ABOVE: Pat Pierce, statistician, supported several scientists including (BELOW) Dr. Peter Layde, who established the Marshfield Epidemiology Study Area (MESA). That became an important source of many epidemiologic research studies.

and other research projects in a community-based setting. Laboratory results, hospitalizations and outpatient visits for each individual registered are captured on an ongoing basis.

It was Dr. Layde, with the assistance of Pat Pierce, a biostatistician, who built the infrastructure for a formal Department of Epidemiology at the Research Foundation with MESA as its cornerstone. The Center grew rapidly and attracted a number of reputable and experienced epidemiologists from various institutions in the country and has become an invaluable resource providing support to the other Centers of the Foundation.

In 1987, Barbara Lee, R.N., who had been a head nurse at St. Joseph's Hospital, joined the Foundation as the first full-time employee of the National Farm Medicine Center, and quickly developed a keen interest in agricultural (farm) injuries occurring in children. In 1995 she received her Ph.D. from the University of Wisconsin-Milwaukee. Her interest in childhood farm injuries continued, and in 1997 she received a $3.5 million grant from the National Institute for Occupational Safety and Health (NIOSH) for the development of a Center of Excellence for Childhood Agricultural Injury Prevention. This large center grant would be renewed in 2003 and in 2008 with assistance from Barbara Marlenga, R.N., Ph.D., who joined the Foundation in 1995.

In 1989, groundbreaking for the Lawton Center occurred just north of the Clinic building. The Lawton Center for Research and Education would be a 55,000 square-foot facility that would house Marshfield Medical Research Foundation, Department of Medical Education and National Farm Medicine Center. The second floor would provide much needed laboratory space. The Center was named after the late Ben Lawton, M.D., who died in 1987. He was a tireless thoracic surgeon and an intrepid leader of the Clinic whose dedication and philosophical wisdom helped guide the Clinic's growth over his 30-year presence. The dedication ceremony took place on July 13, 1990, and the dedication address was given by the Honorable Louis W. Sullivan, M.D., U.S. Secretary of Health and Human Services. Other dignitaries in attendance included The Reverend Thomas Finucan, who gave the invocation and who was also a member of the Research Foundation advisory board; Tommy G. Thompson, then governor of Wisconsin; Robert F. Froehlke, who was the former Secretary of the Army and a staunch supporter of the Research Foundation; and Mrs. Ruth Lawton, widow of the late Dr. Ben Lawton.

ABOVE: The dedication of the new Lawton Center for Research and Education with state and national dignitaries in attendance.

LEFT: U.S. Secretary of Health and Human Services Louis W. Sullivan, M.D. is shown giving the key note address at the Lawton Center dedication ceremony. Ruth Lawton, widow of Dr. Lawton, cutting the ribbon to open the new center.

Also in 1990, Marshfield Medical Research Foundation and Marshfield Clinic merged after much discussion and deliberation on the implications of the merger. Dr. Sautter was then the director of the Research Foundation and Wenzel was the executive director of Marshfield Clinic. The officers of the Foundation were: Richard A. Reinhart, M.D., Chairman-Executive Committee; Douglas J. Reding, M.D., MPH, Vice Chairman; Patrick J. Juneau, Secretary; and Kenneth E. Heiting, Treasurer.

In October, 1990, the Foundation received a four-year grant from NIOSH that was a cooperative agreement with two primary goals: a) undertake an epidemiologic surveillance study to ascertain the unique barriers Wisconsin farmers face in obtaining cancer screening and treatment; and b) provide information on cancer control education, screening and interventional material specifically tailored to the farm community.

In 1991, Dr. Jim Weber was the recipient of the Sebold Award and in 1992 he became director of the Center for Medical Genetics. He methodically added staff to assist in his genetics research projects, and published numerous papers on his research.

Ben Lawton M.D., surgeon and pillar of the Marshfield Clinic for 33 years. Dr. Lawton served as Clinic president on seven different occasions in the years 1969-1979. He also served as chairman of the Board of Regents of the University of Wisconsin (Madison).

In 1991, Paul Gunderson, Ph.D., joined the National Farm Medicine Center from the University of Minnesota where he was in the Department of Health Statistics. His expertise and experience in the agriculture industry and rural health became an important asset to the National Farm Medicine Center (NFMC) and he became director the following year. Also in 1991, Dr. George Magnin retired as the program director for the internal medicine residency training program and Norman Desbiens, M.D.,

Dr. Doug Reding secured a NIOSH grant in 1990 to study cancer in farmers and was principal investigator of the 14-year PLCO grant and several other NCI-funded studies.

was named his successor. Later that year Dr. Joseph Mazza became the director of Medical Education and he established a new and accredited Med-Peds residency training program in 1996. The program was the first of its kind in the state and provided four years of training in primary care, i.e., internal medicine and pediatrics. This program would complement the already existing Internal Medicine and Pediatric residency programs at the Clinic and Saint Joseph's hospital.

In that same year the Foundation received a large 14-year multi-center grant from the National Cancer Institute (NCI), a division of NIH, to study the value of screening procedures for various cancers: prostate, lung, colorectal and ovarian (PLCO). The grant has been renewed and the project continues to this day. Thus far, the Foundation has received more than $21 million to continue the study. The principal investigator was Dr. Douglas Reding, a Marshfield Clinic oncologist.

Dr. Sautter resigned as director of the Research Foundation in December 1993 and Dr. Paul Gunderson was appointed interim director. In 1994 he became the director of Marshfield Medical Research Foundation and remained in that position until April 2000, when he retired from the Foundation and returned to his farm in North Dakota.

In November of 1994 construction of the Melvin R. Laird Center for Research was announced and it would be dedicated in September 1997.

Barbara Lee, RN Ph.D., joined the Foundation in 1987 and became director of National Farm Medicine Center in 2000. Her research on childhood farm injuries has cast her as an international figure related to these issues.

In 1995 Edward Belongia, M.D., joined the Research Foundation as an epidemiologist and became the director of that Center which had grown to six research scientists and approximately 25 support staff. Over the past decade and a half, the number of epidemiologic research projects, grants and contracts escalated, providing the Foundation with cutting-edge information and collaboration with other nationally recognized academic medical centers and government agencies, such as the CDC. Dr. Belongia's research has focused on vaccine safety and effectiveness and in 2004 he received a $1 million grant from CDC to study the effectiveness of influenza vaccines. He has been involved in the evaluation and safety issues of a number of other vaccines in the recent past and is currently the principal investigator for the Vaccine Safety Data link, a collaborative project involving eight research organizations across the U.S. The network provides timely safety information to guide regulatory action by the Food & Drug Administration.

In December of 2000, Robert Greenlee, Ph.D. joined the Epidemiology Center. He updated the MESA database and organized the Foundation's successful bid for affiliation as a full member in the Cancer Research Network, funded by NCI. He became a member of the HMO Research Network Governing Board, and conducts research sponsored by NCI, the National Heart, Lung and Blood Institute, and the Agency for Healthcare Research and Quality. He is an affiliate investigator in the PLCO cancer screening trial and has served as the statistical advisor to the Wisconsin Medical Journal. Through an adjunct appointment, Dr. Greenlee also serves as vice-chair for the Graduate Program in Clinical Investigation at the University of Wisconsin Institute for Clinical and Translational Research.

Paul Gunderson, Ph.D., joined the National Farm Medicine Center (NFMC) in 1991 with research expertise in agricultural safety and health. In 1992 he was named NFMC director and a year later became interim MCRF director. Dr. Gunderson served as MCRF director from 1994 until his retirement in 1999.

An anonymous one million dollar donation for research launched the establishment of the Clinical Research Center in 1998. Its first director was Kurt Reed, M.D., Ph.D., who remained as the Center's director until 2005 when he was succeeded by Steven Yale, M.D. The Center bridges clinical staff of the Marshfield Clinic system with the resources of the Research Foundation. The Center supports and helps design clinical trials, provides assistance in gathering and extracting important data for analysis. It also serves as a vital path to publications on clinical research projects emanating from the institution and serves as a link to advanced medical technology.

Epidemiologist Ed Belongia, M.D. and Gerald Ogdahl, computer specialist, look over the shoulder of Barbara Bartkowiak, Reference Librarian as she demonstrates literature searches during Clinical Research Day.

The new Laird Center for Medical Research was announced in 1994 and dedicated in 1997.

In September 1997, dedication of the Melvin R. Laird Center for Research and Education occurred with a host of national and state dignitaries attending in honor of Melvin Laird. Former President Gerald R. Ford and Former Secretary of State Henry Kissinger were in attendance along with Wisconsin Governor Tommy Thompson, former Wisconsin Governor and U.S. Senator Gaylord Nelson, and U.S. Congressman David Obey. The Center would become the new home for the Genetics Center, the NFMC, the George E. Magnin Medical Library and the Froehlke Auditorium. The building was joined to the Lawton Center and the rest of the campus by a skywalk extending over Kalsched Street.

In February 1998, Dr. Weber's Wisconsin Genetics Project was approved by the Institutional Review Board (IRB) and consisted of a large-scale study of human disease for the purpose of improving healthcare by a better understanding of disease at the molecular level. As Dr. Weber said, "Genetics is the most powerful way of unraveling the molecular biology of disease." The Wisconsin Genetics Project would be published in medical journals and ultimately make its way into hospitals and clinics throughout Wisconsin, the nation and the world.

Ed Belongia, M.D. joined MCRF in 1995 and became Director of Marshfield Epidemiology Research Center (MERC) in 2001. His flu vaccine research achieved international recognition.

Marshfield's Center for Medical Genetics became one of the most highly respected genetics laboratories in the world. It was awarded a $2.4 million contract expansion from the National Heart, Lung and Blood Institute (NHLBI) in 1998. This, along with a $3.9 million contract award in 1994 for mapping genes associated with disease, provided Dr. Weber with the financial support and incentive to expand his laboratory. In 1999 the center was awarded the largest contract in the history of Marshfield Clinic/Marshfield Medical Research Foundation, that being a seven-year, $22.6 million contract extension from the NHLBI over a 12-year period of time.

The most significant achievement of the scientists working in Dr. Weber's laboratory came in 2001 with publication of the Human Genomic Sequence issue of the prestigious scientific journal Nature. This work was carried out by several members of Dr. Weber's lab: Adong Yu, Ph.D.; Chengfeng Zhao, Ph.D.; and Ying Fan, Ph.D.. This paper would be studied and cited by many scientists worldwide. During his tenure at the Foundation, Dr. Weber discovered single nucleotide polymorphisms (SNPs), a small genetic change or

Former U.S. President Gerald Ford and Melvin Laird attend the Melvin R. Laird Center for Research and Education dedication in 1997.

variation that can occur within a person's DNA sequence, that would provide a major breakthrough in the way genetics research would be conducted. As a result of his genetics research, Weber was able to acquire three patents on his work that have become useful tools in virtually every genetics research center in the world.

Robert (Bob) Froehlke played a major role in the acquisition of financial resources that enabled the Foundation to expand and accommodate the increasing number of scientists attracted to the growing reputation of research in Marshfield. He led the fundraising campaign for Laird Center for Research & Education. The building houses the Robert F. Froehlke Auditorium/ Conference Center and is an ongoing tribute to Mr. Froehlke.

In 1999, John Olson, M.D., succeeded Dr. Mazza as director of Medical Education, and forged a closer, more formal relationship with the UW School of Medicine in Madison that resulted in Marshfield Clinic and Research Foundation being designated as the Northern Campus of the Medical School. This was largely due to the growing number of students and programs established to provide primary care experience within the Marshfield Clinic system. Dr. Olson designed and directed these programs at various regional centers of the Clinic in rural northern and central Wisconsin.

Dr. Jim Weber, director of the Genetics Research Center joined the Foundation in 1986, established the Wisconsin Genetics Project, and was awarded the largest contract in the Foundation's history.

CHAPTER 4

The New Millenium

The beginning of the New Millenium was highlighted by the acquisition of more scientists and administrators. This expansion of resources set in motion a number of cutting-edge research programs that were to set the stage for regional and national collaborative efforts in genetics, epidemiology, bioinformatics and clinical medicine. The newly constructed Laird Centers for Medical Research (1997 and 2008) provided accommodating space for additional scientists, laboratories and support staff in the Foundation.

In November 1999, Michael Caldwell, M.D., Ph.D., became director of Marshfield Medical Research Foundation. He had been professor of surgery and biochemistry at the University of Minnesota and was also the founder and director of the Center for Wound Healing and Reparative Medicine at that institution. Dr. Caldwell received his medical degree from the University of South Carolina and his Ph.D. in physiology from Vanderbilt University. During his tenure as director of the Research Foundation, he fostered growth of research support systems, implemented a tenure-track system for scientists' promotion, encouraged many new extramurally-funded grant applications, and established a wound healing clinic which continues to the present.

In addition to laboratory and research space, the Laird Center for Medical Research contains the Froehlke Auditorium, George E. Magnin Medical Library, and the Erdman Lobby.

It was Dr. Caldwell who proposed and launched the Personalized Medicine Research Project (PMRP). This huge endeavor was established using the existing resources of MESA, the Clinic's electronic medical records (EMR) and the capacity of Dr. Weber's mammalian genotyping program. He secured funding to expand the repository of DNA from individuals in central Wisconsin to study various aspects of each individual's genetic profile as it related to their response and risk to various medications. In November of 2000, Steve Wesbrook, Ph.D., under contract to the Research Foundation, was hired as a consultant for PMRP. In June of 2001 he became the associate director of the PMRP program and played an essential role in its development. Dr. Wesbrook wrote the strategic plan and

Dr. Michael D. Caldwell, M.D., Ph.D., vascular surgeon, became director of the Foundation in 1999 and established the Personalized Medicine Research Project.

synchronized the detailed operational planning for this comprehensive initiative. He also assisted the Foundation in communicating the program's vision, goals, and plans by helping to secure $2.8 million in grants and awards and additional financial support for the Foundation. He left his position at the Foundation in November of 2005 but returned to become deputy director of the Research Foundation in November of 2006.

The PMRP biobank is among the largest population-based genetic research projects in the United States. It consists of nearly 20,000 DNA extractions and storage of plasma and serum samples from individuals who have volunteered to be part of the research project. The short-term goal of the project is to establish a database to allow research in epigenetics, pharmacogenetics, and population genetics. The ultimate goal is to translate genetic data into specific knowledge about disease that is clinically relevant and will improve patient care. The database also contains extensive phenotypic and environmental information on each individual registered and became a major component of subsequent, major genetics initiatives.

Events of 2000 included the death of Dr. Sautter. He was recognized and eulogized for his many contributions and accomplishments, as well as his progressive thinking as administrator of the Foundation. His surgical talents are also noteworthy in that he introduced cardiac surgery to Marshfield Clinic. Later that year, Dr. Barbara Lee became director of the NFMC.

In March 2003, a new organizational structure was adopted consistent with the Clinic's three-part mission to serve patients through accessible, high quality health care, research and education. The Foundation's name

Alana Ziaya and the professional staff at the medical library are a major asset to the Foundation and serve the research and medical personnel as well as the students and housestaff.

was changed to Marshfield Clinic Research Foundation and the Marshfield Clinic Education Foundation was created. The newly formed Division of Education administratively brought together the many educational functions of Marshfield Clinic including postgraduate residencies and fellowships, medical student training programs, continuing medical education, the medical library, patient education, nutrition services, corporate education and media services.

Dr. Kurt Reed looking over the monkey pox images from the electron microscope with CNN's Winnie Dunbar.

Later in 2003, Kurt Reed, M.D., Ph.D., director of the Clinical Research Center and a pathologist with a specialty in microbiology, identified a strange viral illness called monkeypox in a young girl from Medford, Wisconsin. The girl had been referred to Marshfield Clinic for further evaluation of strange pox-like lesions on her extremities. It was Dr. Reed in microbiology, along with Erik Stratman, M.D., and John Melski, M.D., dermatologists at the Clinic, who evaluated and made the diagnosis of monkeypox, which was confirmed by laboratory studies and the CDC.

In 2004 a publication on the human genome, co-authored by Dr. Jim Weber, was acknowledged in the prestigious medical journal, The Lancet, as its "Paper of the Year for 2003."

In 2004, Richard Leer, M.D., became the interim director of the Research Foundation while a search for a new director to succeed Dr. Caldwell was undertaken. Dr. Caldwell left the position and returned to the Department of Vascular Surgery where he continued directing the wound healing program. Dr. Leer would continue as interim director of the Research Foundation until June 2005.

"In 2003 the Marshfield Clinic Education Foundation was created to be consistent with the Clinic's 3-part mission of patient care, research and education."

In April 2005 Humberto Vidaillet, M.D., a Clinic cardiologist, was named director of the Research Foundation. Dr. Vidaillet received his medical degree from the University of Oklahoma and trained at Mayo Clinic and Duke University Medical Center. He was heavily immersed in research activities within the Marshfield Clinic system when he became director. That same year, Dr. Steve Yale was named the director of the Clinical Research Center.

Sebold Award Winners

Gwen D. Sebold Research Fellowship Award given by D. David (Dewey) Sebold. "The $5,000 Sebold award recognizes an outstanding Marshfield medical complex researcher and supports continued research in his or her chosen field. This award was established in 1987 by D. David Sebold in honor of his sister, Gwen, a longtime Marshfield Clinic stenographer." (She died of cancer in 1974.)

1988 **Richard Sautter, M.D.**
1989 **Dean Emanuel, M.D.**
1990 **Duane Tewksbury, Ph.D.**
1991 **James Weber, Ph.D.**
1992 **Thomas Nikolai, M.D.**
1993 **William Myers, M.D.**
1994 **W. Bruce Fye, M.D.**
1995 **Norman Desbiens, M.D.**
1996 **James Burmester, Ph.D.**
1997 **Kurt Reed, M.D.**
1998 **Douglas Reding, M.D.**
1999 **Paul Gunderson, Ph.D.**
2000 **Dean Stueland, M.D.**
2001 **Daniel Jacobson, M.D.**
2002 **Humberto Vidaillet, M.D.**
2003 **Barbara Lee, Ph.D.**
2004 **Edward A. Belongia, M.D., Ph.D.**
2005 **Joseph J. Mazza, M.D.**
2006 **Catherine McCarty, Ph.D.**
2007 **Tarit K. Banerjee, M.D.**
2008 **Loren A. Rolak, M.D.**
2009 **Barbara L. Marlenga, Ph.D.**
2010 **Percy N. Karanjia, M.D.**
2011 **Robert T. Greenlee, Ph.D., MPH**
2012 **Michael Caldwell, M.D., Ph.D.**

Catherine McCarty Ph.D., assumed interim directorship for the Center for Human Genetics in 2004 and became principal investigator of PMRP in 2006.

Dr. Weber left the Research Foundation to begin his own company in Marshfield, Prevention Genetics. Dr. Weber would continue to work with the Foundation as an adjunct professor and consultant in genetics research. As a result of Dr. Weber's resignation, Catherine McCarty, Ph.D., was named interim director of the Center for Human Genetics in 2004. She became principal investigator of PMRP in January 2006, and continued in that position until 2009.

Also in 2005, the Foundation began discussions with Justin Starren M.D., Ph.D., at Columbia University in New York City, in the interest of establishing a bioinformatics research center. In August 2006, Dr. Starren joined the Research Foundation as director of the newly established Bioinformatics Research Center (BIRC). The history of this Center actually dates to 1999-2000 when Carl Christensen, Peg Peissig and Tom Berg, working in the Clinic's Information System Department, saw the need for a designated center capitalizing on research opportunities associated with medical information. This concept was encouraged by others who recognized the importance of being able to house the existing data repositories into one center for research purposes. It was Christensen who then sought and recruited Dr. Starren and served as the interim director of the new Center before Starren arrived in 2006. The BIRC programs would be driven by state-of-the-art computational technology and complemented by the Clinic's data warehouse, electronic medical record, and MESA. By July 1, 2009 there were 32 individuals working in the BIRC.

The Erdman Lobby donor wall is a testament to the many individuals, businesses and foundations who provided support for the Laird Center for Medical Research.

In April 2006, the National Farm Medicine Center celebrated its 25th anniversary with a special event that included testimonials by a number of national media and political leaders as well as local farmers who had benefited from and worked with NFMC through the years.

In September 2006, the Foundation broke ground on the new Laird Center for Medical Research, just north of the existing Laird Center. This would be an 112,300 square-foot, $40 million addition to the Laird Center for Medical Research.

In April 2007, Dr. Erik Stratman was named director of Medical Education. He had been active in teaching both students and residents in the Department of Dermatology, and was committed to the work of Dr. John Olson, his predecessor, in continuing to foster stronger relationships with the University of Wisconsin in the further development and expansion of student programs, and in promoting Continuing Medical Education (CME) and Graduate Medical Education programs. He had also been active on a national level and recently named chairman of the Council on Education for the American Academy of Dermatology.

A formalized collaborative agreement was reached with the University of Wisconsin leading to a major federal grant for research infrastructure.

In September 2007, during Dr. Vidaillet's directorship of the Foundation, a formalized collaborative agreement was reached with the University of Wisconsin to seek a federally funded Clinical Translational Science Award (CTSA) by combining the resources existing at the research centers of both institutions. This would lead to numerous collaborative research projects and grant submissions involving UW and Foundation staff. Numerous CTSAs have been established throughout the country, led by the National Center for Research Resources, forming a national consortium of institutions that would be linked together to pursue clinical and translational science and generate preliminary data for submission of grant applications. The importance of the partnership with the University of Wisconsin Health Sciences Schools was quickly realized by both institutions,

Elias A. Zerhouni M.D., Director of the National Institutes of Health gave the Dedication address for the second Laird Center in 2008.

Humberto Vidaillet M.D., cardiologist and researcher, became director of the Foundation in July of 2005. He linked the Foundation to the UW Institute for Clinical and Translational Research.

overcoming 98 years of coexistence 120 miles apart and a history of competition in clinical research with effectively no collaboration. Dr. Mark Drezner, director of the Institution's Center for Translational Research (ICTR), and principal investigator on the CTSA grant proposal at the UW, inserted complimentary statements in the grant proposal, acknowledging Marshfield Clinic's role. "The historical dedication of the Marshfield Clinic to achieve excellence in translational research and engaging in practice-based clinical interventional medicine, was of paramount importance to the community engagement efforts that were a central focus. Indeed, the population available for study by the two institutions spans a wide range of social and demographic characteristics representative of communities throughout the United States."

In October 2008, dedication of the second Laird Center for Medical Research occurred. This world-class facility was intended to bring together comprehensive patient care and research at Marshfield Clinic Research Foundation. Elias A. Zerhouni, M.D., director of the NIH, gave the dedication address. Other dignitaries and political leaders at the ceremony included U.S. Congressman David Obey, Wisconsin Governor James Doyle and the Honorable Melvin R. Laird. The facility houses the Department of Pathology, clinical and research laboratories, and BIRC.

In 2008, the Wisconsin Genomics Initiative (WGI) was developed as a collaborative research project involving Marshfield Clinic Research Foundation, the Medical College of Wisconsin, the University of Wisconsin-Madison and the University of Wisconsin-Milwaukee. The WGI was state funded initially and would link the resources of each institution and form the proposed "Research Triangle of Wisconsin", i.e. Marshfield, Madison and Milwaukee. With the advances in molecular genetics and epigenetics over the past decade, scientists would be able to predict an individual's risk of developing a disease, target a personalized therapeutic approach and ultimately prevent disease before it occurs. Governor Doyle, who had been promulgating the initiative, pointed to the magnitude and importance of the project and said, "With our combined knowledge, expertise and technologies here in Wisconsin, we have an incredible opportunity to become a worldwide leader in personalized health care. By aligning the intellectual capital of four major institutions, we will meet an important scientific and public health need that could otherwise not be met, and which cannot be accomplished anywhere else but Wisconsin." The Foundation was excited and looking forward to its role in this endeavor.

The Many Names of the **Marshfield Clinic Research Foundation**

- **[19]59** — Marshfield Clinic Foundation, for Medical Research and Education, Inc. (from State of Wisonsin charter September 14, 1959)
- **1968** — "Dr. Dean A. Emanuel was elected the new president of the Marshfield Clinic Foundation for Medical Education and Research" (from 'Cliniscope' Vol. 1, #1, January 1968)
- **1967** — 1967 - 1973 Annual Reports = The Marshfield Clinic Foundation For Medical Research and Education
- **1973** — The Marshfield Clinic Foundation for Medical Research and Education (from 'Pulse' June 15, 1973)
- **1974** — 1974 - 1978 Annual Reports = Marshfield Medical Foundation Inc.
- **1981** — 1981 Annual Report = Marshfield Medical Foundation
- **1985** — Marshfield Medical Research Foundation (from the 1985 Annual Report – "a new name, a new look...")
- **1999** — Marshfield Medical Research and Education Foundation (from the 1999 Marshfield Clinic System Review 'Walking the Talk')
- **2001** — Marshfield Medical Research Foundation (from 2001 System Review 'The whole is greater than the sum of its parts')
- **2003** — "We will begin to implement our name change to Marshfield Clinic Research Foundation on Monday, March 3, 2003" (from an email sent to all Foundation employees from Steve Wesbrook, Ph.D.)
- **2004** — Marshfield Clinic Research Foundation (from 2004 Marshfield Clinic System Review 'Where the future of medicine lives')

In addition to the chronology of the listed events and accomplishments that have occurred over the past 50 years, education has remained an important piece of the Foundation's mission. The expansion of the education programs over the past half century is noteworthy and additional programs continue to be added as the Clinic and Foundation pursue their missions. Evidence of the growing list of new education programs can be confirmed by the following: the Division of Education has had a psychology fellowship program in place since 1992; a three-year dermatology residency that includes both inpatient and outpatient environments was established in 2004; and a hospice and palliative care fellowship was begun in 2003. Additionally, the Foundation has for more than 40 years sponsored a summer student program for undergraduate students, which has attracted applicants from colleges and universities throughout the country. The program provides a research experience in the various centers and laboratories within the medical complex.

Research and education have grown together and will continue as long as we uphold our commitment to the mission statement of the Institution. Tangible evidence of that commitment is the publications authored by many researchers and clinicians who have participated in major, multi-centered studies, including:

- ALLHAT (Antihypertensive and Lipid-Lowering Treatment to Prevent Heart Attack Trial) study, an international trial that looked at the use of antihypertensive and lipid-lowering agents in the treatment to prevent heart disease with the principal investigator being Richard A. Dart, M.D. from the Department of Nephrology.

- AFFIRM (Atrial Fibrillation Follow-Up Investigation of Rhythm Management) study followed patients with atrial fibrillation to investigate cardiac rhythm management. This was a large, multi-center trial that demonstrated that atrial fibrillation can be treated effectively with anticoagulation, but the efficacy of anti-arrhythmic drugs remains uncertain. The principal investigator was Dr. Humberto Vidaillet.

- SUPPORT (Study to Understand Prognoses and Preferences for Outcomes and Risks of Treatments) study was a controlled observational study aimed at improving end-of-life decision making and reducing the frequency of mechanically supported, painful and prolonged process of dying in a large cohort of seriously ill patients. This was a two-year trial with five major teaching hospitals in the U.S. participating. Investigators included Norman Desbiens, M.D.; Peter Layde, M.D.; Marilyn Follen, R.N., M.S.N; and Dr. Vidaillet.

- Stroke Prevention study, which was sponsored by the National Institution of Neurological Disorders, looked at selected individuals who were at risk for stroke from narrowing of their carotid artery. This was a seven-year clinical trial that investigated the effectiveness of surgical intervention (carotid endarterectomy) in reducing the incidence of stroke in this group of patients. The principal investigators of this study were Percy Karanjia, M.D., and Kenneth Madden, M.D., both from the Department of Neurology, and Mark Swanson, M.D., a vascular surgeon.

These were just a few of the significant research studies impacting directly on medical practice standards in which clinicians and support staff have been involved under the aegis of the Foundation.

In 2008, there were 160 publications from Foundation scientists and 61 research projects supported by grants and contracts. There were 17 industry sponsored clinical trials initiated. Additionally, the Foundation sponsors and supports two peer reviewed journals that are recognized and listed by the National Library of Medicine: Clinical Medicine & Research and Journal of Agromedicine.
As of 2009, the Research Foundation consists of five major components or centers. The Clinical Research Center, Biomedical Informatics Research Center, the Center for Human Genetics, the Epidemiology Research Center, and the National Farm Medicine Center. These Centers are bound to the mission of the Foundation and exchange resources with one another to maximize research opportunities. Each has a significant degree of autonomy

with respect to its function and agenda and communicates with the director and deputy director of the Foundation. There are of approximately 30 Ph.D. and M.D. investigators and more than 160 support staff. In addition, there are approximately 150 physicians and other healthcare professionals in the Marshfield Clinic system engaged in research. The spectrum of services available that are essential to research include the Core Lab, Biostatistics and Bioinformatics, Research Integrity and Protection (the Institutional Review Board committee), Scientific Writing and Publications, Sponsored Programs, Fiscal Affairs, and Health Economics.

In addition to the five Centers listed above, the Foundation has always had a rich history and tradition of research in infectious disease. This can be traced to the work of Dr. Dean Emanuel, Fritz Wenzel and others in their search for the etiology of Farmer's Lung and other chronic respiratory diseases during the earlier decades of the Foundation. Therefore, it is appropriate that we acknowledge the accomplishments of the individuals who have been and are involved in infectious disease research at the Foundation: The diagnosis of monkeypox in 2003 by Drs. Kurt Reed, Erik Stratman and John Melski; the ongoing vaccine trials that have been directed by Ray Haselby, D.O., and Bradley Sullivan, M.D.; the published work of Jennifer Meece, Ph.D., and her colleagues on mosquito-borne viral infections and blastomycosis; and the search for water-borne pathogens

Dean Emanuel M.D., cardiologist, researcher and founder of the National Farm Medicine Center is revered as the "Father of Agromedicine."

Education Day at the medical center where students, residents and medical staff have gathered in attendance for the presentations.

and water contamination in the lab of Mark Borchardt, Ph.D. Additionally, the Foundation has become a national participant in infectious diseases with cutting edge research on the genetics and epidemiology of methicillin-resistant Staphylococcus aureus (MRSA) through the work of Sanjay Shukla, Ph.D., and his staff. Dr. Shukla's laboratory reported two new pathogens in 2003 and in 2005 he was awarded a prestigious RO1 grant from NIH to continue his research on MRSA.

It is important to recognize the many individuals who played such an important philanthropic role in the expansion and success of the Foundation's recent past. The tireless efforts of the Honorable Melvin R. Laird and his wife Carol, and the Honorable Robert F. (Bob) Froehlke and his wife Nancy are appreciated and without their help, much of the philanthropic support received would not have been possible. The naming of the Laird Center for Medical Research and Froehlke Auditorium are testaments to their efforts.

Many individuals, businesses, and foundations have supported Marshfield Clinic Research Foundation over the years and their contributions have nourished the Foundation's growth and accomplishments. We provide special recognition to William D. Connor Jr. and his wife Jane, Delta Dental of Wisconsin, Frank R. & Betty J. Koller, Chester Krause, Lockheed Martin, Marshall Erdman and Associates, Ministry Health Care, The David and Lucile Packard Foundation, Security Health Plan, the DeWitt & Lila Wallace Funds, and the Reader's Digest Fund for their generosity. Last but not least, the creation of the Laird Center for Medical Research would not have been possible without the $11 million grant given by the U.S. Congress in recognition of the Honorable Melvin R. Laird and Honorable John E. Fogarty and their important work to facilitate medical research throughout the United States.

Through the years, the Foundation has been guided by two essential committees: the Board of Trustees and the National Advisory Council. The Board of Trustees meets five times per year and is comprised of 20 members; eight elected physicians from the Clinic Board of Directors, one senior scientist, 10 lay members selected from the community and the Clinic president. The Board of Trustees oversees the activities of the Research Division of Marshfield Clinic and reports to the Clinic's Executive Committee and Board of Directors. The National Advisory Council, formally adopted in 1984, includes influential, successful business and professional individuals who provide guidance on Clinic and Foundation policies and strategic planning.

Conclusion

The history of the Research Foundation has been one of steady growth and success, and has been inextricably tied to Marshfield Clinic. The Research Foundation was founded to "benefit mankind" as stated in its Articles of Incorporation. Today it continues its commitment to improving the quality of life for many individuals through research, education and community service. The accomplishments of the Foundation are legion and have provided a national reputation and recognition resulting in the continuous acquisition of grants and contracts to sustain the many research projects emanating from its scientists. Education programs have expanded to encompass many levels of biomedical disciplines, and the health care plan continues to provide comprehensive services for its many members.

The Foundation's future lies in the commitment of the next generation of scientists and physicians who will inherit this legacy and who will impose their values and talents on the needs that lie ahead in the endless pursuit of mankind's quest for knowledge.

Acknowledgements

I wish to express my gratitude to the many individuals who provided me with the recorded and unrecorded information that reflect the rich history of the Research Foundation.In particular, I want to thank Alana Ziaya and Lyle Hoerter, who are the curators of the archives, the emeritus physicians (Emanuel, Lewis and Magnin) who were always available to set things straight, Fritz Wenzel, who was there from the very beginning, and Marie Fleisner, who transcribed my dictated tapes during the early phases of the report. Shawn Williams of the Creative Services Department turned the narrative into a welcoming history book with placement of photos and illustrations. The constant encouragement of Dr. Barbara Lee kept the project moving forward and she helped verify research facts and captions. And last but hardly least, I am grateful to my wife Ginny who has helped with the syntax, editing and collating of all the materials from the project's inception through its completion.

Made in the USA
Monee, IL
05 November 2026

67382721R10036